Money

Physics and Distributive Justice

Oded Kafri

Money
Physics and
Distributive Justice

Oded Kafri

Production by: Mendele Electronic Books Ltd

Translation from Hebrew: Yitzhak Solsky

"Everything is foreseen, yet free will is given"

Mishna-Tractate Avot 3:15 A.R. Rabbi Akiva

Table of Contents

Preface   **9**

Chapter 1: The Characteristics of Money   **17**

Money in the Affluent Society   17

Money as a Fluid   29

Chapter 2: Physics and Economics   **33**

Money as Energy – the Gas Economy   33

The Problematics of Gas Economy   36

Money as Heat   39

Economic Networks and the Heat Economy   43

Characteristics of Networks   49

The Statistics of Heat – the Planck-Benford Distribution   55

The Planck-Benford Law – History and Uses   61

Chapter 3: The Statistics of Money   **70**

Constraints and economic inequality   70

The Gini Index   88

Poverty and Wealth   101

Executive Salaries   112

The Distribution of Wealth – the Pareto Principle   124

Chapter 4: Boom and Bust   **129**

Summary   **139**

Bibliography   **143**

Preface

In which the contents of the book are presented in a nutshell. Do not be scared of a bit of physics; you may skip it and still understand the idea.

If you ask anybody, "What is money?" they are most likely to look at you with wonder, thinking, "What does this smart aleck want from me?" It is strange, though, that in modern society—in which our lives revolve around money—there is no answer to this basic question.

If we look up "money" in, for example, the Merriam-Webster Dictionary, we will find the following definitions: legal tender, a medium of exchange, a measure of value, a standard for deferred payments, etc. These definitions, alongside others of the same type—whatever their meanings—do not explain what money is, but rather describe its uses. Some think that money is a commodity produced by a government in a factory that prints bills and mints coins for its citizens to use for commercial transactions. But the money that the state produces, also known as cash, is only a tiny fraction of the money in public hands.

Today it is already clear that in the near future, the cash we hold in our wallets will disappear. All money will be in banks and we will pay by directly transferring a certain amount of money from our account to a provider's account (or to any other account) using clearing companies, smartphone applications, and other means as they become available through the Internet. All this, without any contact with cash.

In this book we will present money as a physical entity. For this purpose, we will 'fast forward' a bit in time and describe a future world without cash, which will resemble a network with many nodes and links. We will present a simple model of an isolated country that runs an independent economy. We will call that economy an "economic network" and assume that this country has a single central bank which serves as a hub for all commerce. All of the money in the hands of citizens of this country (we will call these "nodes") is stored in this bank.

A node of our network may be a single person or any economic body whatsoever—a limited company, an association, etc.—that has an account with the bank. Thus, when Bob pays M dollars to Alice, -M (minus M) dollars will be registered to Bob's account in the bank, and +M dollars will be registered to Alice's bank account—the actual amount of money in the bank will not change as a result of these payments.

The sum of all debits and credits in a node's account at a given moment is called a "balance." The surprising conclusion that derives from this model is that the total amount of money in our imaginary country is zero. Bob's payment to Alice only changes their respective balances, so one may generalize and say that trade changes the balances in the nodes, but not the aggregate amount of money—the total value within the bank.

One may ask: how do Bob and the other residents of the country have money? The answer: bank loans. Obviously, some lucky fellows have positive balances in their accounts, do not owe money to the bank, and do not need a loan. However, at some point in the chain of payments, someone has taken a loan from the bank so he can, for example, pay his employees' salaries or buy merchandise from suppliers. The node receives a loan from the bank in exchange for collateral offered by the node. This collateral is not money but assets, such as financial or tangible assets—real estate, for instance—which are pledged to the bank. Therefore, there are credits equal in amount to all the debits in the bank

Why does a bank give loans? Because the bank wants to make a profit, and it gains the most profit from the interest that it charges on the loans it provides to various nodes.

At this stage, we should introduce assets[1] into our economic model in more detail. When someone asks the bank for a loan, he has to convince the bank that the amount of the loan will be returned, and will therefore have to demonstrate his ability to repay the loan. For instance, if that person needs a loan in order to purchase an apartment for $300,000, the bank may demand that he pledge the apartment to the bank. The bank may in addition require that person to buy insurance, making the bank a beneficiary of his insurance policy in case he becomes incapacitated. Generally, if the asset is "worth" $300,000, the bank will lend less than its market value, so that if the borrower defaults on the loan, the bank can quickly sell the property—possibly at a loss—and recover as much of its money as possible.

Therefore, while the varying value of assets is the heart of economics, the fact of the matter is that the amount of money in the country is zero. "Money" is nothing but loans moving between nodes of a network.

---

1. The word "assets" denotes a wide variety of products and goods – but here we only refer to tangible and highly negotiable assets.

In the economy of the free world today, a tiny part of the population supplies all our basic needs—i.e. food, clothing and shelter. Therefore, the value of most assets is based on human whims, not objective value, and is determined by changing fashions. Nevertheless, if you tell someone that a painting by a great artist is worth nothing because it lacks objective value, you may find yourself being laughed at before receiving the crushing reply, "Sure, but I just sold it for a million dollars yesterday." If we substitute the word "assets" with the word "payment" we will see that people trade everything: real estate, knowledge, pleasures of the body, decorative objects, lottery cards, and innumerable gadgets. The reasons for trading in assets are varied, sometimes strange, and always subject to change. Things that were extremely valuable become very affordable, and vice versa. For instance, the lobster, used to feed prisoners in the nineteenth century, is today considered an expensive delicacy; while pearls, formerly the very symbol of wealth, have lost their luster. If we assume that goods are only an excuse for trade between nodes, we can make an analogy between our economic network, in which the nodes exchange dollars, and the World Wide Web, in which they exchange bits. To extend the analogy, we may say that a rich man resembles a busy node, while a poor man is a low traffic node.

What is so special about this description of the economy? Well, this description resembles a cluster of material particles which absorb and emit energy particles and are known in physics terminology as "photons in equilibrium." Photons are energy particles that reveal themselves to us as light and heat, and are characterized and defined by their amount of energy and their location. The location that characterizes a photon is called "spatial radiation mode."

A cluster of particles that absorbs and emits photons from the radiation modes in equilibrium, which we will use as an analogy for an "economic network," is a well-known phenomenon in physics and is called "black-body radiation." This phenomenon was explained in 1901 by a German scientist named Max Planck. According to Planck, each object in the universe, including the universe itself, is a "black body." The material particles comprising the black body do not affect the nature of its radiation, just as the nature of goods does not affect an economy. In a black body, regardless of what materials it is made of, photons are absorbed and emitted. Photons are completely different from the particles of tangible matter: the particles of tangible matter collide with each other, while photons are "transparent" to each other and therefore do not collide. As a consequence of this strange nature, there can be an infinite amount of them in the same location, just as there is no limit to the amount of money one may place in a single bank account.

Photons always move at the speed of light, which has a fixed value in a vacuum. We, therefore, notice them only when they are absorbed or emitted, thus transferring their energy into a certain medium—that is to say, to material particles that will emit them to another radiation mode, and to others that will absorb them from another radiation mode. The same goes for money: if we liken a radiation mode to a bank account, photons would then be analogous to money transferred from one account to another for merchandise or for any other recompense, which is analogous to a material particle.

The distribution of photon energy between radiation modes—or, in our analogy, the distribution of money between entities—depends upon a single parameter only: the temperature of the black body, its energy, which is analogous to the wealth of the country. Surprisingly enough, if the number of photons is relatively large compared to the number of radiation modes, the distribution of energy among them is universal—it is independent of the energy of the black body (its temperature) or any other parameter. In our analogy to economics, if the amount of money is much greater than the number of traders (bank accounts) or, alternatively, the levels of income—such as deciles or percentiles—the distribution of money among the residents of the nation is universal and independent of the total wealth of the nation.

The distribution of money among people is one of the key issues of human society. Two thousand years ago, Plutarch, a Greek historian and philosopher, said that "an imbalance between rich and poor is the oldest and most fatal ailment of all republics."[2] It is common knowledge that many of history's revolutions broke out due to feelings of discrimination and inequality among the masses. Later in this book, we will see that economic inequality between people who live in developed economies—the OECD countries—resembles the inequality in energies between the spatial radiation modes of a black body.

The formula used in this book to describe the inequality and the distribution of money between income levels was derived from the Planck distribution known as the Planck-Benford distribution. The Planck-Benford formula is

_____

2. Some claim that this often-quoted maxim was "invented" in the 1960s.

derived from the second law of thermodynamics and states that, in a closed system, the probabilities of all of the system's different states will be such as to bring the number of possible combinations of the system's states to a maximum.[3]

A fascinating fact about the Planck-Benford distribution formula, which is nothing but a statistical distribution, is its universality. If we use it to calculate distribution of wealth, and we subdivide the population into ten levels, it will describe the distribution of wealth between the deciles in any country; if we use it to predict election results, it will describe the distribution of voters between parties; if we use it to calculate the distribution of decimal digits in random files, we will get the frequency of the digits (known as the Benford Law) as well as many other well-known phenomena (known as Zipf's Law or the long tail distribution[4]).

In this book, we examine the model and its implications about the distribution of wealth between people, both intuitively and quantitatively. We look into the inequality of income between deciles, as represented by the Gini Index, and into the percentage of relative poverty.[5] We will calculate, according to the Planck-Benford Law, what the salary of a CEO should be as a function of the

---

3. In economics, the equivalent to a "state" is a bank account. Maximizing the number of combinations of states yields thermodynamic equilibrium. This point will be explained in chapter 2.
4. The concepts mentioned in this paragraph, such as Zipf's Law and the Planck-Benford formula, and those mentioned in the following paragraphs, such as the Gini Index, will be explained later in the book.
5. A poor person is defined as one earning one half or less of the median income.

number of employees in the company and the average salary in it, as well as the percentage of assets that the various percentiles and deciles should possess.

The results obtained are surprisingly similar to those of both the OECD states and major companies, such as those appearing in the Fortune 100 list.[6] One may, therefore, surmise that economics and human behavior are also subject to the second law of thermodynamics and that we, human beings, play a secondary role in the game.

This book is a sequel to *Entropy – God's Dice Game*. The technical material and mathematical derivations are found in scientific publications listed in the bibliography at the end of the book. The book is intended for a general reader who is interested in the nature of things, and for the economists, sociologists, and politicians who shape the economy.

Acknowledgements

I would like to thank Hava Kafri for working closely with me on the book, and our friend Hagit Shapira for her excellent work in editing the book. Special thanks also go to Moshe Pereg for his commentary and insight. Thanks to Yariv Kafri, Ilana Glatt-Brönnimann, Zeev Hochberg, Eli Fishof, Dany Rosenne, Zvi Aloni, Galit Kafri and Yafa Franco who read the book and gave useful remarks. Special thanks to the translator of the book to English, Yitzhak Solsky, for his work and for his useful remarks.

---

6. The list of the topmost-ranking one hundred companies according to Fortune magazine, which ranks international companies by various criteria.

Chapter 1

The Characteristics of Money

In this chapter we will discuss the difference between money and tangible assets; how money's uses in the past differ from its uses in today's affluent society; and money's special characteristics as a fluid entity.

Money in the Affluent Society

It seems that everybody knows what money is, but the concept of "money" has not actually been defined. In this sense, the science of economics is centuries behind the hard sciences. Five hundred years ago, in physics, too, there was no definition of energy—just as today we do not have a definition of money. You obviously should not worry about that if you are a businessman; one may, of course, earn plenty of money without knowing its definition—just as work could be produced through the application of energy before its essence was understood.

Money differs from tangible assets such as objects and goods (e.g. the metals of silver and gold); money also differs from many other physical entities, whose properties are known and measured—or, in technical terminology, entities which are subject to a conservation law.

When someone has a certain amount of money in his bank account—let's say M dollars—this is not the same as saying that he has a house. A house can be used for residence or rent and may be sold—that is to say, the house has a physical existence and multiple possible uses, while money can only be used to pay somebody. A house is a tangible asset, while money is only a record in a bank account.

Until the late twentieth century, money was linked to the value of certain commodities, the best known of which are silver and gold. The USA had a "gold standard" that made it theoretically possible to always exchange about 35 US dollars for an ounce of gold. This standard was revoked in 1971 by then - President Richard Nixon, due to the inflation caused by the vast costs of the Vietnam War.[7] As we will later see, it is impossible to link money to any commodity in modern economics. If we assumed that X tons of gold were kept in the basements of the Federal Reserve Bank (the Fed) of the United States, based on which the bank would issue a certain amount of bills, the amount of money would be subject to a conservation law; however, we know that throughout the economic ebb and flow, the amount of money in public hands undergoes sharp changes. Another difference between money and goods is that goods have an independent existence whereas money requires a bank. Gold will exist without the Fed, but dollars won't. The need for banks is particular to modern societies, in which one requires a vast variety of products in order to exist. One can exist in primitive societies without money even today—but in the modern world, in which you drink French wine, use an American phone that was produced in China, and buy coffee whose

---

7. Indeed, the Gold Standard was already fictitious by that time, as the US forbade individual citizens to possess gold in 1934, after the Great Depression.

beans grew in South America and were roasted in Italy, banks are a necessity. The wealth in people's hands is composed of tangible assets and money, whose quantity is recorded in the bank.

The ratio between liquid cash and assets is the essence of economics. Money is used to transfer ownership of assets and goods between people. We will see that money has certain statistical characteristics that are independent of the value of the assets, but which have to do with the essence of money itself. These important properties are the main subject matter of this book.

Someone once said, "If you know how much you're worth, you're not worth much." It is indeed hard to know how much someone is "worth", as the value of assets changes constantly. The best example is the stock market— the graph reflecting stock prices moves incessantly, like a seismograph needle. In theory, if you have real assets, their worth is mainly a function of the alternatives in the market. If the bank offers you 8% per year for your money, for example, and purchasing an apartment for rental will yield 2%, and if we further assume that one can generate an unlimited amount of apartments, you may prefer to sell your apartment and invest your money in the bank. Nevertheless, massive sales of apartments by investors will bring their prices down, contractors will no longer find it advisable to build apartments, and the result will be a shortage of apartments and a corresponding rise in rental fees. Income from apartments will increase until at some point you will find it advisable to take your money out of the bank and invest it in an apartment again.

The balance between demand and supply is not necessarily a rational phenomenon as the one described in the example above, though. "Hot" goods may lose their public appeal for unclear reasons. Among the best known historical examples is the 17th century obsession with tulips in the Netherlands. In 1636, the price of a tulip bulb in the Netherlands—a global superpower in those days—was more than ten times the annual salary of a competent professional, equivalent to approximately half a million dollars in today's money. Then, in early 1637, the fad wore off and the price of tulip bulbs plummeted dramatically in the absence of demand, losing almost all value in no time flat. Nowadays that period is known as the "Tulip Mania."

Israel experienced similar bubbles; one example is the trade of gold coins produced by the Israel Coins and Medals Corporation in the early 1970s. I remember a gold coin that weighed about half an ounce. I bought it in 1972 for 420 Israeli pounds—about half an engineer's monthly salary at the time—and sold it for 2,700 Israeli pounds some three months later. I immediately used the proceeds to buy three gold medals, each weighing a full ounce. Now, more than 40 years have passed and Israeli coins and medals have lost their appeal. Their monetary value is now mainly determined by the amount of gold inside them. In the end, I received only a negligible value for the transaction as a whole, even though at the time I bought the medals I had multiplied the worth of the initial coin by a factor of about six and a half.

The uncertainty of the value of assets is the essence of economics. If we leave perishable goods aside, the expectation of an increase in the value of an asset raises its price, while the expectation for a decrease in its value lowers it. Thus the present value of an asset is dictated by people's expectations of its future price.

Why do we need money to subsist? In the not-too-distant past, in isolated communities such as early-twentieth-century Israel (then Palestine, or "Palestine—Land of Israel" in Hebrew), in the days of the pioneers, people mainly consumed basic commodities, most of which were produced on-site. Money was not that important. For example, my father owned a farm, on which he raised cows and chickens. He also had a field in which he grew food for them, as well as a small orchard. We ate eggs laid by our hens, cheese prepared by my mother from the cows' milk, as well as chicken meat (when a hen fell ill). My parents sold the surplus to the neighbors and bought bread at the bakery across the street. I'm sure that my father could obtain fabric in return for my mother's cheese, for her to order a dress from the dressmaker. My father had a well and a septic tank that drained the sewage of the farm and the family. The family was almost perfectly economically independent. It only depended—like all farmers do—upon the graces of nature. Indeed, in many cases people did not use the government tender issued by the British mandate[8], preferring instead to issue their own small promissory notes that could be used to buy bread, milk and similar products at the grocery store.

---

8. The Egyptian Pound was used in Israel (then Palestine) until 1927. It was then replaced by a Palestinian pound.

*Grocery money from the 1920s
(the author's collection)[9].*

A translation of the above.

How did my father obtain his land? Since the village was remote and plagued by malaria, the land was relatively cheap. He worked as a laborer, ate only oranges and bread for several years, saved his money, and bought the land from the communal committee. He built the house he lived in from bricks that he made himself. He made extra bricks, and exchanged them for shingles. This kind of life—in a remote, provincial society—shaped my father's economic views. In his old age, watching a

---

9. "Mil" was one thousandth of an Israeli pound until 1960—translator's note.

musical performance on TV, he wondered, "If everybody dances and sings, who does the work?"

Everybody today knows that a song is a much better product than an orange. A huge economic gap lies between the orange and the song: the orange can be sold once, while the song can be sold millions of times. An orange is a tangible product; a song is information.

In the village where my father lived during the early twentieth century, it was hard to make a living out of information. Residents consumed information through newspapers (one person would buy the newspaper and five others would take turns reading it) and books. Nobody in the village produced information. Today, most of my acquaintances produce information: computer software, entertainment, blogs, banking, commerce, etc. And how about agricultural goods, energy, and other tangible products? In a way, these are also mainly made out of information. Due to the increasing sophistication of production methods, knowledge is the main component of almost every product. For example, the seeds used by farmers are industrial products in every sense, carefully bred or genetically engineered, and their price derives from the R&D costs that went into their production—the knowledge to produce the thing, not the thing itself, has value.

Huge changes in technology and communications have created a situation where a tiny percentage of mankind produces the breadbasket of a century ago with a fraction of the labor. An independent, largely self-sufficient farm, such as my father had, could not exist today. In exchange for taxes, authorities provide infrastructure, deliver water, drain wastewater, build roads, plan the environment, and more. Nobody may raise cows and hens or produce cheese without endless licenses. The production of food has been regulated and streamlined. One needs well-developed transportation facilities and available communication to reach the workplace. Most of our expenses—both on essential products and services and on luxuries—go into products and services that did not exist in the past. As a result of technological developments, the prices of raw materials are on a permanent declining trend. A British soldier earned one silver pound per month in the eighteenth century. Today a pound of silver costs less than $250, while that soldier earns almost $3,000 per month. We can see that in real terms, silver has lost nearly 90% of its value. The price of oil continues to decrease, against all predictions, due to new production methods. Commerce is becoming increasingly web-based and the price of i.e. mobile phones is mainly a function of the price of the knowledge used in their production.

In the early 1970s, a multi-disciplinary group of experts met in Europe in order to predict the upper limit to humanity's growth.[10] The study was motivated by a resonating argument: as the number of people grows, and as the amount of raw materials available is finite,

10. The Limits to Growth—a nonfiction book written in 1972 by Donella Meadows, Denise Meadows, Jørgen Randers and William W. Behrens III. The book was commissioned by the Club of Rome. Using a computer-based model, it describes the ratio between economic growth and increasing populations—and the limitations of natural resources and growing pollution. This book was translated to 30 languages and sold 30 million copies. It is perhaps the most widely known book in the field of environmental sciences.

obviously the population of Earth cannot keep growing indefinitely. How much, then, can it grow? After lengthy discussions, those experts predicted that within fifteen years, which is to say by the late 1980s, most metals would have run out entirely, including iron, aluminum, and copper, as well as petroleum and other raw materials. They concluded that growth could not continue for any extended period. It is worth mentioning that all of these raw materials are very common in nature and the problem was that they were difficult to produce. For instance, it was once so expensive to produce aluminum out of the mineral aluminum oxide[11] that, in the late nineteenth century, aluminum was used as a substitute for silver in luxury cutlery and tableware. In those years, window frames in Israeli homes were made of wood; nowadays most window frames are made of aluminum, and wooden frames are more expensive, wood being a natural material that cannot be industrially produced. Not only had aluminum not disappeared, it had become an extremely cheap material as the result of new developments in the knowledge necessary to produce it, leading to a great increase in its use. A similar situation exists in the fuel market: as the result of new developments in drilling technology, the real price of petroleum saw a marked decrease. A similar situation can be seen with many other raw materials as well.

We can see, then, that the real goods—such as metals or foods—have no fixed value. Now let us ask: is the total value of assets subject to a conservation law? When Bob gives Alice an amount of money (M), it is agreed

---

11. Aluminum oxide is among the most common minerals on Earth.

that Bob has lost M money while Alice has gained M money. This approach is known as a zero-sum game. If the world of economics behaved that way, the value of all of the assets in the world would be fixed. Let us remember, though, that between 2000 and the present (2017), the world underwent two economic crises: the dot-com bubble[12] in 2001 and the US subprime crisis in 2007, which still has the world licking its wounds a decade after its inception. Trillions of dollars were written off of the wealth of the entire population in both crises. Furthermore, the US government and the governments of Europe apply what is known as an "expansionary monetary policy." In colloquial terms, we say that the treasury prints vast amounts of money (although money is not actually issued in a printing house but, as we will see later, is created by giving cheap government credit). One wonders—if money is a commodity, as economists are wont to say, and it is true that when one floods the market with a commodity its price goes down, then would it not mean that the real prices of other commodities should have risen? This phenomenon, this decrease in the value of money, is called "inflation." But prices do not always go up, and deflation even exists in some places—prices go down as the result of a rise in the value of money.

After the technology stock crisis that saw prices plummet to less than half their former value, I was asked by a friend—a physicist, who is used to the fact that physical quantities are conserved—"Where did the money disappear to?" He was sure that some band of crooks, somewhere overseas, had stolen it. The simple

---

12. The crash of internet stocks in the NASDAQ stock exchange to about one-half of their value in 2000–2001.

answer is that the money never existed. Assets did, and they remained; the thing that disappeared was a considerable portion of their worth.

How can this be, given that actual money left investors' pockets and went to stock exchanges? This marvel can be explained using the following thought experiment: let us say that a town has 10,000 identical houses. Let us then say that each house is priced at $100,000. What is the value of the real property in the town? If a house is sold for that price whenever a buyer appears, the real property is worth $1 billion. However, if everybody in the town wishes to go elsewhere, and if everybody wishes to sell their houses at the same time, and if no demand exists for those houses, the houses in the town are worth nothing. If only one person lived in the whole world—meaning that, theoretically, the entire Earth were his property—he would be destitute. This is because the very essence of the value of money is in giving it to somebody else and receiving something in exchange. The recompense may be real, such as an orange or a massage; intangible, such as listening to a song, watching a movie, or using software; or conceptual, where the recompense is money or stocks. Let us say that I gave $100 to someone, and in exchange, he gave me EUR 95, or a share of a certain company. In both cases, I gave some paper and received other pieces of paper. Obviously the price of the paper itself is negligible in this transaction, and what actually happened was that a registry operation took place in the bank to denote that I have $100 less and that I am to be credited with EUR 95, a certain corporate share, or another conceptual property such as electronic money, which is itself practically another kind of money that can exist without regulation.[13]

A share of a public company that is traded in one of the stock exchanges of the world reflects, supposedly, the worth of that company divided by the existing number of shares. Had we issued shares in the imaginary town from our experiment, and written in the stock exchange that the town had 10,000 shares, a share would be priced at $100,000 and the town would be worth $1 billion. Unlike an apartment, which is a real asset in that its owner may use it (live in it or rent it out), the share can only be traded. If you have a Google share, for instance, the only way you can use it is to sell it – provided, of course, that there are buyers. On the other hand, if you happen to pass by the company's headquarters and feel like using its restrooms (which are known for their perfect cleanliness), or buying lunch in its subsidized cafeteria (which pleases the eye and offers fine flavors), we may assume that the guard at the gates will not let you in, even if you explain to him that you are among the owners of the company.[14]

---

13. The Bitcoin is an example of a virtual, international currency that is freely traded on the Internet. It is a conceptual asset, similar to a stock or a foreign currency. Electronic money differs from other types of conceptual assets in that anyone can produce it through some complex computer work (this action is called 'mining'). The Bitcoin is consequently not subject to any type of regulation, which is why some countries do not allow trading in it. In contradistinction to money that is produced by a bank's loans against collaterals, the virtual coins are not backed by any assets and therefore their value is purely speculative.

14. It's worth mentioning that some companies distribute a portion of their profits as dividends, which is a different case.

So why do people purchase shares? This is an excellent question indeed. If the company does not distribute dividends, the value of the share is exactly like the value of the Dutch tulips. In fact, even a Picasso painting that was recently sold for $170 million is a conceptual asset whose worth is based on the assumption that somebody, for some unclear reason, will be willing to pay even more money for the right to keep it in an air-conditioned cellar and protect it from harm. Obviously, anyone who purchases a conceptual asset expects its price to rise for some reason. Just as there is no clear reason for the initial price, there is no clear reason for the price to increase. What difference does it make to you how much Google makes in profits, or if it will have higher profits next year, if the profit is not shared with you?

Thus, if most of the essence of the financial market is as unexplainable a phenomenon as the Tulip Mania, why does it exist?

Money as a Fluid

We will now argue that unlike numerous material assets, money always flows in ways that we do not control. We cannot make money disappear or create it—we can only guide it. For example, let's see if we can remove money from the capital market. Let us say that you have $100,000 that you wish to keep in the bank as savings. You open a savings account and let us say, for the sake of simplicity, that the bank offers you no interest—but you can withdraw the money anytime you want. Your money does not stay in the bank, of course. The bank lends the money out to other clients and buys financial and other assets with it, at the bank's discretion. For the safety of the depositors, the state demands capital adequacy from the bank—or, in simple terms, that the amount of money in the bank must be a certain percentage, say 10%, of the total value of its investments. This does not mean that the bank must hold 10% of the value of its clients' deposits in a safe, but that it deposit the money with a "safer" establishment, such as the state. The state spends the money on the military, on police, on regulations; on rewarding its voters, its employees, and its elected officials; and on certain more esoteric investments that a private investor would never consider making, out of fear that he might find himself involuntarily committed by order of the district psychiatrist.

Let us assume now that the bank runs into difficulties and asks for its money back from the state. Obviously, the government will not sell the army or stop paying salaries to those it employs—it will "request" that the treasury of the state transfer money to the bank and demand that the bank re-organize. *De facto*, the act of transferring the funds to the bank is a registry operation performed in the bank by the state, to the effect that the bank may keep operating without hindrance despite its difficulties and despite not maintaining capital adequacy—as long as it cuts, for example, the number of its employees.

We see, then, that when we deposit our money in the bank, our money does not really leave the capital market but remains in the general money bathtub.

Let us now try a different way of removing money from the market. For example, let's buy a golden nugget and keep it under our bed. Obviously, this act did not change the amount of gold or the amount of money in the world.

In the not-too-distant future, almost all payments will be nothing but registry operations in the bank—Bob transferring M dollars to Alice. In a world in which payments become solely records of bank transfers, we may describe the economy as a single giant web, in which any two nodes (such as people who are connected through the internet) transfer or do not transfer monies among themselves.

This begs the question: what about the goods? We argue that goods are essential to the economy, but their essence is unimportant. The sentence "the essence of the goods is unimportant" sounds absurd, admittedly. It is obvious that nobody will pay money to someone without receiving some kind of actual recompense. But we will have to agree that other than the basic products that we require for our existence (which are produced by only a tiny fraction of the population in modern society, and those people can barely make a living), we pay money for things that we desire without an objective reason. Why do we desire those things? Because they are regarded as coveted items in the social network that we belong to—see the matter of the tulip bulbs and Picasso's paintings.

Many men would pay plenty of money to have hair grow on their heads. Why? Because for some reason, which may change from time to time, hair on one's head is regarded as handsome. Another example is the earlobe. If somebody loses his earlobe, he runs to a plastic surgeon to produce an earlobe for him. Why? Because everybody has one.

In summary: the amount of money in public hands has no fixed value. Increases (deflation) or decreases (inflation) in the value of money are unexpected and therefore unpredictable. The value of money may rise in periods in which credit is given for free and fall when credit is expensive. At times, when markets trend downwards, the worldwide value of assets diminishes. When the markets trend upwards, value increases.

Unlike tangible goods that may be isolated—for example, placing an orange in the refrigerator—one cannot isolate money. Money always flows.

It may appear that money as an entity is totally removed from the world of physics; that it is a conceptual entity invented by Man, that omnipotent being. This conceptual entity is not subject to any conservation law and its behavior is unexpected, unpredictable. We will further see, however, that money is a physical entity and that its behavior is subject to the laws of nature.

Chapter 2

Physics and Economics

In this chapter, we will present two approaches to modeling the economy and its driving power—money—in physical terms. One approach, the "gas economy," treats money as energy, while the other considers money as heat. We will delineate the limitations of the first approach, whose conclusions are inconsistent with actual economic reality, and show the advantages of the second approach.

Money as Energy—the Gas Economy

Attempts have been made to describe the field of economics using physical laws since the 1990s—this area of research is known as econophysics. A salient attempt was made to find a correlation between the economy and a gas (Dragulescu & Yakovcnko, 2000; Ksenzhek, 2007).

Consider a closed, isolated gas tank that contains a fixed number of molecules and a fixed amount of energy. If we liken a gas tank to an isolated country and a molecule to a human being, the molecule's energy is its money. The molecules keep colliding with each other all the time, exchanging energy among themselves, which is to say that they "trade" with energy. We would, therefore, expect that whenever we inject energy into a gas container, the gas will heat up—i.e., the molecules will collide more frequently and the average amount of energy in each molecule will increase. Injecting energy into the tank is equivalent, in that case, to the state injecting money into the economy.

The distribution of energy among the molecules in a gas in equilibrium[15] has a bell-like curve. This distribution, known as the Maxwell-Boltzmann Distribution, also describes, among many other phenomena, the distribution of height among people—with the difference, though, that a person's height is approximately fixed throughout that person's adult life (excluding the growth period and old age).

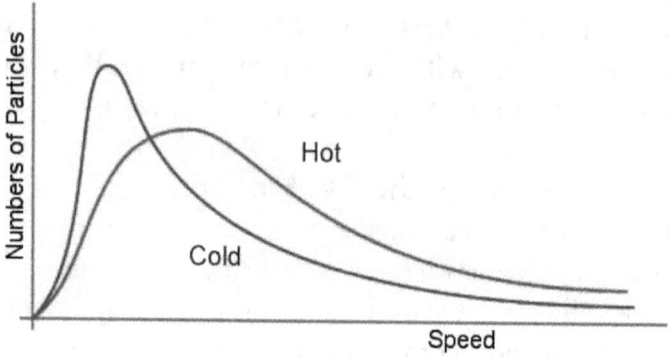

Figure 1: The Maxwell-Boltzmann Curve, showing the relative number of gas particles as a function of their speed, is a bell-shaped curve. Its highest point is at the average speed. The hotter the gas, the lower the curve and the higher the average speed.

Conversely, the individual energy of the gas molecules is subject to constant changes with each collision, exactly like the amount of money held by a specific individual. Nevertheless, the distribution of energy among the

15. Equilibrium is a stable state in which the entropy of a system is at its maximum value.

molecules remains constant, depending only upon the total amount of energy stored in the gas. "Trade" in a gas is done through collisions, in which some molecules transfer energy to some other molecules.

If we subdivide a population into ten deciles, each having the same number of people and all people within a decile having identical incomes, and analyze the distribution of income among deciles using the gas economy as a model, we get the graph presented in figure 2.

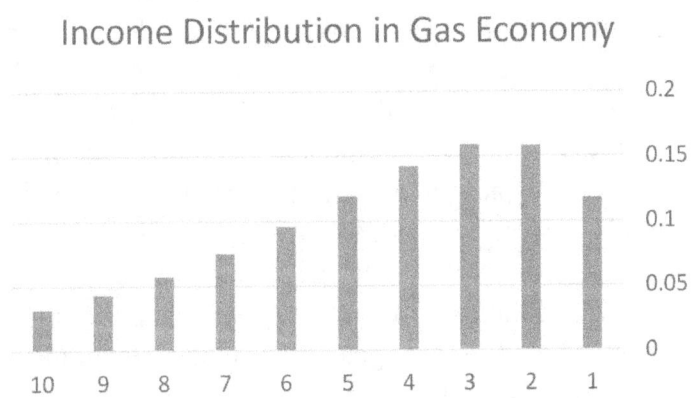

Figure 2: The distribution of relative income by deciles in a gas-like economy. The curve shown here varies as a function of the total amount of money and is not fixed; see figure 1.

Had income been equal in all deciles, each decile would gain one-tenth of the overall income. In figure 2 we see that the third decile is the richest—about five times more so than the poorest decile.

This curve shows a greater number of molecules whose energy is close to the average than one would find in a realistic economy. It implies that most people have money around the national average—relatively few are poor and relatively few are rich. It further implies that it would be as difficult to find a rich person whose wealth is triple the average as it would be to find a person whose height is triple the average.

The Problematics of the Gas Economy

The model presented in the previous section, which compared money to energy, has a number of problems:

- In reality, the distribution of wealth among people is not similar to the Maxwell-Boltzmann bell curve distribution of energy. Most people earn less than the average. The distribution of money among people, known as the 'long tail distribution,' is unequal: there are very few people who have tens of billions of dollars available and many people whose available funds only allow them to survive very frugally. The gas economy is relatively egalitarian in comparison to the real economy, in which the wealth gaps are much bigger. Moreover, the gas economy is a deterministic economy—that is to say, if we know the amount of energy in the tank (the total wealth) and the number of gas particles (the number of people), we can precisely determine its future course of development, which is toward equilibrium with a known average wealth. This is in contrast to a real economy, which has better and worse periods and a fluctuating average level of wealth.

- When a molecule leaves a gas tank through a tiny hole, it leaves together with its energy. In a real economy, money has no meaning without additional inter-trading nodes. A rich man who fills his suitcase with money, buys a boat, and reaches an uninhabited island immediately becomes destitute because he cannot trade with his money—unlike a molecule, which can be rich in energy all on its own.

• Imagine that molecules could multiply like unicellular organisms. The molecules in the gas tank would now share their energy with more molecules and therefore each molecule would have less energy. The analogy to human society implies that when more people have to share the same resources, they will become poorer—but reality shows that such intuitions, as expressed by gloomy economic forecasts since a long time ago, are completely mistaken and that the opposite is true. The main problem with such economic forecasts is an assumption based on the following "iron-clad logic": as the amount of goods is fixed, the number of people who can live on them is limited. This was the conclusion of the "Limits to Growth" report, which predicted that very soon raw materials would run out, oxygen would disappear, forests would be uprooted, and humanity would become extinct. Is that so? The facts show that despite such predictions—and in complete contradiction of the gas economy model—even though the global population has grown immensely, the standard of living has risen, accompanied by a better life expectancy.

More recently, futurists have predicted that, as only a few people were going to produce everything we would need, the others would not have to work. They suggested cultivating a leisure culture so that the non-working classes could get a good tan, read, travel, etc. Obviously and regretfully this is not the way things work. In order to convince the farmer to give away the food he produces and the mason the houses he builds, one has to give something in return—perhaps a song, or a piece of furniture, or some wondrous device, and so on.

As a result, the amount of goods constantly increases. In addition, the modern way of life brings new basic commodities into existence, such as transportation, computers, the Internet, and telecommunications.

- A gas contained in a closed tank has a fixed amount of energy. In economics, this would mean that money would have its own conservation law, similar to the conservation of energy.

Now, what would happen if we connected a few gas tanks and allowed the molecules in all tanks to bring their average energy to a common level? Indeed, the second law of thermodynamics would cause the gas molecules in the various tanks to average out. The energy would flow from the hotter tanks to the cooler ones.

Now, let's imagine a number of countries, with varying levels of wealth and varying numbers of residents, which trade with each other. Would this same phenomenon take place? It does, in fact, happen, and money does flow from richer to poorer countries. It seems self-evident that energy would flow from a hot tank to a cold one, but it is not so clear why money should flow from the rich to the poor. The sociological reason is that the poor do many (mostly unimportant) things for the rich, thereby convincing them to buy those things for money. We know the story about the Spanish sailors who reached South America and exchanged mirrors and shiny beads with the Indians for gold. Actually, the Spanish sailors exchanged with the Indians one folly for another. The Indians, being isolated from the world, didn't know that the gold folly had a bigger market than the shiny pieces of glass, which could easily be mass-produced in Europe. Nowadays we buy cheap products from relatively poor China, where prices are low, in exchange for money, which we send there from the richer West. Money will keep flowing to China until the average amount of money in China is equal to that in the United States, for instance. In physical terms, we may say that the energy from the richer West will flow into poor China until the per-capita level of energy in both countries is the same.

Theoretically, since the future course of the economy is expected, we could predict its future behavior. In the Last Days, the average incomes of all people in all countries would be the same. Additionally, since the economy is a zero-sum game, there will be no severe economic crises, in which everyone loses, and no 'bubbles' will be created. There will always be some who will gain and some who will lose. This picture seems quite idyllic, except for the nagging point that each person in it will have an inherent interest in there being fewer people, because obviously when you have fewer molecules in any tank, more energy is left for the remainder.

Money as Heat

So far, we have seen that the gas economy is not realistic. There is no law of "conservation of money" to parallel conservation of energy. Money cannot be isolated in the same way as energy, as in a capacitor or a battery. Likewise, nobody may remove money from an economic network in the same way as energy may be removed from a thermal reservoir, as when one removes an energy-rich molecule from a gas container.

We will now claim that money is analogous to **heat** rather than energy.

It may appear that heat and energy are the same entity, as they are measured by the same physical units. But **heat is energy transferred between two bodies.** Even though it is generally said that our lives revolve around the energy that we consume, one of the indexes of our standard of living being the amount of energy that we consume, we actually only use heat. Most of the energy that feeds Earth is the heat that reaches us from the relatively hot sun to the comparatively cool Earth. We operate the electric appliances in our homes using energy that we receive from the electric company, for which we pay according to our consumption. We consume information from the radio, TV, and Internet thanks to suppliers who broadcast energy to our receivers. Most importantly, we communicate by transferring energy from one person to another, using electromagnetic waves for seeing, acoustic waves for hearing, and electromagnetic oscillations that move at the speed of light for phone and Internet communications. Heat is an energy that flows from a hot body to a cold one, according to the second law of thermodynamics.

In and of itself, matter is also energy—multifaceted energy: the energy stored in its mass, the energy of its movement (kinetic energy), the energy related to forces exerted on it (potential energy), and so on.

Heat, on the other hand, is mostly waves of pure energy. The waves move at an immense speed—about 300,000 kilometers per second (the speed of light). Heat particles have no mass and therefore neither kinetic nor potential energy.

Heat particles, called photons[16], have an important characteristic that distinguishes them from material particles: unlike material particles, which occupy space and therefore cannot exist in the same space at the same time as one another, heat particles can be located in the same space without any limitation.[17] The location characteristic of a specific photon is called radiation spatial mode.

Since many photons can occupy the same volume of space, their statistical distribution is completely different from that of a gas. The energy distribution of a gas, as predicted by the Maxwell-Boltzmann curve, resembles a bell; whereas the distribution of heat energy between radiation modes (the Planck Distribution) produces, under certain conditions that also occur in multiple social phenomena, the long tail distribution. Figure 3 below shows this in the distribution of income among deciles. We can see that the top decile (marked 1), which has the highest income levels, earns about seven times more than the tenth decile (marked 10). We also see that the income differences between the lower deciles are small, while the income difference between the two top deciles is large.

---

16. By way of analogy, there are also sound waves named "phonons," which move at the speed of sound, and also gravitational waves. We will not discuss these here.

17. Strictly speaking, in special laboratories, physicists do manage to cram several atoms of certain liquids, such as helium, into the same volume of space for short time periods.

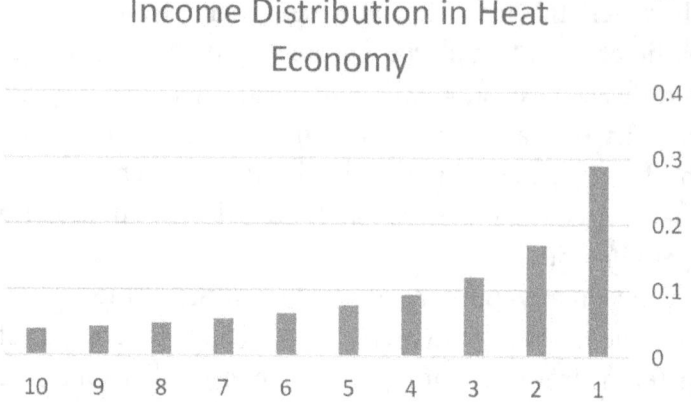

Figure 3: The relative distribution of income between deciles in a heat economy.

Figure 3 shows the distribution of income among deciles in a heat economy. When discussing the gas economy (figure 2), we assumed that money was distributed among people like energy among molecules; in heat distribution (figure 3), we assume that money is distributed among people like the energy of photons among their modes. In other words, we compare the physical system, in which photon energy is transferred between various modes, to an economic system, in which money is transferred between bank accounts. The photons need material particles to absorb and emit them in order to go from one mode to another, and in the same way, money requires recompense to move from one account to another.

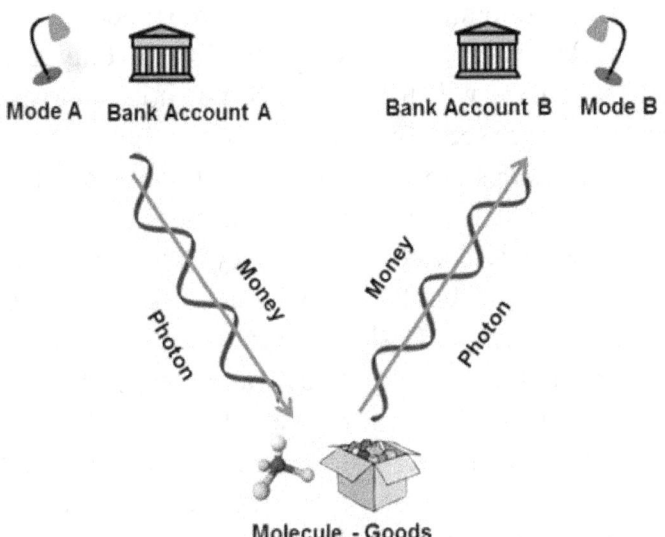

Diagram 1: The analogy between absorption and emission of a photon by a molecule, and a bank transfer between accounts in exchange for goods

The set of bank accounts between which money transfers are carried out as goods change hands between account owners is an economic network. This network is analogous to that of the modes that exchange photons (heat) by absorbing and emitting them to (and from) material particles. We will now describe an economic network, and later discuss general characteristics of networks.

Economic Networks and the Heat Economy
A general discussion of networks could fill many volumes with material that does not concern us here. We will, therefore, focus on a simple model of a network economy, upon which we will be able to project the distribution of

energy among the photon modes, in order to analyze the statistical-economic characteristics of the model. We will mainly elaborate on income distribution in equilibrium, from which we will try to deduce potential courses of development for economies.

This model describes a group of people who exclusively use bank transfers when trading. Those people buy and sell goods and services with money, which they have in their bank accounts. We will call the bank's total clients a "country". Cash and barter deals do not exist in this country. When anybody purchases anything for amount M, a "-M" entry is registered in his bank account (payment is taken out) and the seller is credited with a "+M" entry. The recompense (merchandise, services, or conceptual property) is the essential reason for the money transfer, but the nature of that recompense is unimportant, as today's "hot" product may lose its entire value by tomorrow. All money transfers between clients are registered in the bank. The sum of all of the balances of the all of the bank's clients is the total amount of money in the country.

How was the existing money in the country created? Surprisingly, for money to be created, the bank must allow certain clients to have a negative balance. When the bank lets a client do this, we say that an individual has received credit from the bank. When the bank gives credit, a payment (-) of the credit amount (C) is registered in the balance sheets of the bank. The person receiving the credit pays the money he receives to his suppliers and employees, who now have revenues registered in their accounts, in the total amount of C. The aggregate amount of money in public hands is the same as the aggregate amount of their balances, and this money obviously originates in the bank. Therefore, the balance of money in public hands is equal to the total debts owed by people to the bank.

To illustrate this model, here's a detailed example: a man named Tycoonie requests a large credit from the bank, claiming that he would use it to manufacture a product that he would be able to sell to others for an amount that greatly exceeds the credit amount. If the bank were a voluntary institution, it would answer Tycoonie with a resounding "NO"—the bank has no interest in helping Tycoonie enrich himself with other people's money. But if Tycoonie offers a commission to the bank—a certain percentage of the credit that he will receive for the loan period—the bank will earn money, and so will find it advisable to grant the requested credit to Tycoonie. This commission is called interest (I). Let's now assume that Tycoonie does well in his affairs—the bank gave C money to Tycoonie, who used C to pay his suppliers and employees, who then deposited C in the bank. In addition, Tycoonie, whose venture was successful, made a profit from his clients (P). Tycoonie paid back C value to the bank, plus I value in interest—and now his account shows an amount of (P-I).

Let's check the accounting in this process:

- Tycoonie's clients paid him (C+P)—this amount was deducted from their balances.
- The suppliers received C from Tycoonie—this amount was credited to them.
- Tycoonie had (P-I) registered to his credit.
- The bank had I registered to its credit.

In this process, money only changed hands—just like in the story of two merchants who took a walk and suddenly saw a frog. Merchant A says to merchant B, "Give me a hundred dollars and I will swallow the frog." B agrees, so A swallows the frog and receives the hundred. B immediately regrets his rashness and wishes he had not wasted the hundred dollars—he now suggests swallowing a frog himself in return for his money back. A agrees and the deal takes place. After a while, A says to B, "We both swallowed frogs and nothing changed," to which B answers "Yes, but we had turnover!"

Things are not always quite so rosy, though. What happens when Tycoonie borrows the amount of C from the bank and then loses it? In many cases, the bank has to write off Tycoonie's debts. Let us now check the accounts:

- The suppliers registered +C in the bank.
- Tycoonie registered –C in the bank.

Unfortunately, though, Tycoonie cannot repay his debt. Therefore, ignoring the accrued interest:

- The bank registered "0" in Tycoonie's account.
- The bank registered "-C" to itself.

If the bank does not go bankrupt and can still accommodate its clients' money withdrawals, then there is no problem. However, if the bank cannot give people their money back, a problem is created and the system has to be restarted. The system is restarted by either nationalizing the bank—meaning that all of the residents of the country pay for the deposits of the bank's clients—or by having another bank buy the bank and reach an arrangement for the benefit of creditors in which the bank's clients will receive a part of their money. This procedure is known as a "haircut"—just like the barbershop where they cut your hair, only here they cut your wealth.

In order to prevent banks from going bankrupt, as occurred in Iceland in 2008[18], countries make it a point to have a larger number of banks and regulate those banks in the hope of preventing this from happening. Nowadays we know that the US government was wrong to have allowed the Lehman Brothers Bank to reach bankruptcy during the US Subprime Crisis in 2008.[19] The fall of Lehman Brothers was seen by many as one of the main reasons for the economic crisis described in the summary.

Consider another scenario. Tycoonie takes the loan from the bank, invests it abroad, and loses his money there. When Tycoonie loses his money in his own country, the bank's money goes to Tycoonie's suppliers, who deposit it back in the bank; but when he loses his money abroad, the money disappears from his country, thereby reducing the amount of money available for trade in that country. Under free market conditions, such losses are offset by investments made in Tycoonie's country by tycoons from other countries. Foreign investments are desirable for a country's economy. Regardless, though, the overall amount of money in the world was not affected by Tycoonie's losses.

In the above model, we see that money does not behave like energy but rather like heat. Money is created or disappears through registration transactions at the bank, in much the same way photons are emitted and absorbed

---

18. All three major banks in Iceland went bankrupt and were nationalized in 2008.
19. Lehman Brothers, an American-based global holdings corporation, filed a creditor protection claim after finding itself in debt for about $613 billion. This was the biggest bankruptcy in history.

between modes in a black body. This process actually also resembles the transfer of information between nodes on the Internet. If we try to draw preliminary conclusions from this simple model, we can see that the total amount of money in the country is zero. A debt is registered to someone for each amount of money transferred between the clients of the bank. The bank seems to have collateral for the credit it gives—such assets as real property, warranties, rights, etc.—but it is clear that the value of those assets is unfixed and could disappear at any moment.

Well, how much money is transferred among the bank's clients, then? The answer to that depends. On what? On the "mood" of the public as demonstrated by their willingness to take loans. When the public's willingness to take loans from the bank grows we call it a boom, and when the willingness diminishes we call it a bust. This can be compared to Bipolar Disorder – a disorder that causes extreme mood swings. One manifestation of these swings is that when a patient is in a manic period, he tends to spend money, while during a depressive phase he becomes a miser. The economy behaves in a similar way—when the desire for credit grows, people in that country are happy; when it diminishes, they are depressed. How is the national mood generated? The reasons for that are unpredictable. We can say that if half of the population is happy, and the other half is sad—a simple statistical prediction—then, as we will explain later, members in the network tend to join the majority and think like the majority. Therefore, small fluctuations in this balance of half happy, half sad may cause a great boom or a great bust. To understand economics better, let us describe some general characteristics of networks.

Characteristics of Networks
The following are some useful things to know about the physics behind economic networks. This section is written to be accessible to the general reader, but may be skipped if it is too difficult.

In the heat economy network, the bank accounts are nodes and money is heat flowing among the nodes. Therefore, in order to create a quantitative model of a heat economy, we must first understand why and how heat flows.

The nature of heat was first investigated by a French engineer named Sadi Carnot. Carnot's study—which was published in 1824 and in which he explained the concept of heat and its relation to mechanical work for the first time—was done in order to find out the maximum amount of work that could be produced by a steam engine.[20] A steam engine is a heat machine that produces mechanical work out of the energy transferred between two different temperatures—one high, the other low—between which heat flows. In a steam engine, the high temperature—about a hundred degrees centigrade—is the temperature of the steam, and the low temperature is the room temperature of the machine's surroundings.

Carnot found that the maximum efficiency of a steam engine, or of any other kind of heat-based machine, depended exclusively upon those two temperatures—the high one and the low one. It is for this reason that the maximum efficiency obtainable from a heat machine is called the "Carnot efficiency." The Carnot efficiency is

---

20. James Watt invented the steam engine, the Industrial Revolution generator, in the late eighteenth century.

one of the definitions of what was later called "the second law of thermodynamics." About forty years after Carnot, in 1867, a Prussian named Rudolph Clausius showed that the Carnot efficiency was a law of nature. He showed that the Carnot efficiency is equivalent to the assumption that the sum of the amount of heat, Q_i, transferred in any process, then divided by the temperatures of the emitters or absorbers, T_i, is a quantity that always increases in irreversible processes, does not change in reversible processes, and therefore never decreases. Clausius called the quantity Q/T "entropy." This leads to a simple wording for the second law of thermodynamics, as follows: ***The entropy of the universe always increases in irreversible processes and never decreases.***

The inclination for irreversible changes in nature, partly due to the irreversible nature of time itself, dictates that entropy should grow and grow. Until when? Until the system under observation reaches equilibrium. Why does entropy not grow any further upon reaching equilibrium? Probably because it has then reached its maximum extent.

Within about twenty years of Clausius' work, several physicists—Ludwig Boltzmann (Austria), Josiah Gibbs (USA) and Max Planck (Germany)—showed that the manifestation of entropy in equilibrium is probability-based, reflecting the maximum possible number of different states in which a system may be found. For example, given an economy made of three trade arenas, the greatest amount of different states in the system will occur when one arena holds about 50% of the trade, a second arena about 29%, and the third arena about 21% of trade. This distribution applies to other areas as well. If, in a certain country, there are three political parties, votes will be distributed (in equilibrium) between the parties as follows: 50%, 29%, 21%.

Obviously, the second law does not tell us which arena or which party will receive 50%—but it does predict the relative distribution of the quantities in equilibrium. The same law, with slightly different assumptions, predicts a bell-like curve (similar to figure 1) for intelligence, longevity and many other quantities. The known laws of probability actually originate from the second law of thermodynamics (Kafri, 2016).

Therefore, the statistical interpretation of the second law also makes it possible to calculate, using the same method, the distribution of links among the nodes of a network in equilibrium.

The term "network" describes various types of entities. We should elaborate on the differences between an economic network (which resembles the Internet) and most other networks. We can differentiate here between two types of network: static networks and dynamic networks. In **static networks**, the link between any two nodes is static, or fixed. There are N nodes and they are interconnected by L links. Many static networks are created in nature spontaneously, such as the network of rivers, the network of branches in a tree, or the network of blood vessels in our bodies. Man also creates static networks, such as the water piping network, the sewage piping network, the train network, and the electrical network.

Are networks created by humans any different than those spontaneously generated by nature? The answer is no. If we believe that man-made works differ from any other spontaneous creation of nature, we will have to accept the hypothesis that Man is a supreme creature, capable of acting above or against the laws of nature. This hypothesis would be tantamount to discarding the entire scientific method.

All of these networks are designed for transportation. The railway network, for instance, is designed to transport passengers. Those travelling from station A to station B may board a train that travels at regular times, during which the railway between those two stations is closed for other trains to prevent collisions. Of course, we want to avoid collisions because the train and the passengers are made of material particles, which cannot be in the same space at the same time. **Switched networks** are those in which, when a certain body goes through a certain route, traffic in that route is blocked for other bodies. When a connection between station A and station B is desired, a switch must be thrown. Similarly, when we want water, we open the tap, which is itself a switch that allows water to flow from the water network to a glass. When we want to illuminate a space we throw the light switch, etc. The math describing such networks is mainly geometry-based.[21]

Dynamic networks, whose main real-world representative is the Internet, are those without switches. Theoretically, all nodes are always interconnected and traffic between them is not regulated by switches—all of the links are always open and traffic is always possible between all nodes. The Internet, which is in essence an energy flow, differs from traffic in a fully switched network in that it corresponds to the statistics of **heat**, unlike switched networks, which have statistics more akin

---

21. Graph theory.

to **matter**. It seems that the statistical characteristics of non-switched networks, such as those of social networks, predict totally different economic behavior than those predicted by the gas economy.

Before examining additional differences between switched and non-switched networks, though, we will define the quantities that characterize an economic network.

The first quantity in the network is its **number of nodes**. If we describe the economy of a country as a network, the number of nodes denotes the number of traders. In other words, the number of people, companies, NGOs, or any other associations that have a bank account—in short, all of the bank accounts.

The second quantity is the **number of links**. In a static network, links are physical and fixed, like the rails between two train stations. In an economic network, the link is the registration of a money transfer in the bank—a simultaneous debit and credit—and therefore a transient event. The link is usually caused by the receipt of certain recompense by one node from another. Nodes are also connected physically at times—but this is not essential, just as it is not essential for a person to receive recompense for his money.

Links, which are carried out by bank computers at the speed of light, have a quantity reflecting the amount of money transferred. Unlike the train network, through which only one person may be transferred at a time—as only one person may occupy each seat—many transactions may be registered at the bank simultaneously, regardless of their size: one does not have to transfer each dollar in turn, but can rather register the entire transferred amount at once.

If we connect all of the absolute values of the transactions, i.e. the total value of the links, then divide the sum by the total number of nodes, the resulting ratio (W) represents the average economic activity of a node in a network.

Since in any money transfer one node is registered (+M) and another node is registered (-M), W is not subject to any conservation law and may increase or decrease according to the amount of credit consumed by the nodes and provided by the bank.

Energy flows among the nodes because energy flow increases entropy, per the second law of thermodynamics. Similarly, the flow of money also increases entropy in an economic system, and therefore the system tends towards equilibrium according to the second law of thermodynamics.

Entropy is proportional to the logarithm of the number of different states in which a statistical system may exist. It is a measure of the uncertainty of that system (Kafri & Kafri, 2013). Similarly, in an economic network the number of different states is the number of different possibilities for money transfers among bank accounts. The entropy of the economic network, then, is a measure of the uncertainty of that economy.

Let us now describe some network characteristics that derive from the second law of thermodynamics.[22]

As a rule, the probability of an event occurring increases as the entropy generated by the event increases. That is to say, when two different events may occur, the event that would generate more entropy is more likely to occur than the event that would generate less entropy (even though

22. Explanations and mathematical proofs may be found in literature (for example Kafri & Kafri, 2013).

there is always a possibility of either event occurring). The ratio between the probabilities of both events will be that which would bring the overall entropy in the system to a maximum. Therefore, the greater the number of different possible events, the greater the increase in the total entropy. The entropy generated by the plurality of possible events is called the "mixing entropy." If we call a network a "country," nodes "people," and the average money transfer per person "the wealth of the country," we can show, using the laws of physics, that:

1. If one has to choose between relocating to a rich country with a high W, or to a poor country with a low W, he is likelier to relocate to the rich country (Kafri, 2014).

2. Similarly, money is more likely to flow from a rich country to a poor one than vice versa (Kafri, 2014a).

Both of these rules lead to the conclusion that the population of rich countries tends to grow, while the wealth of rich countries tends to diminish. When will these processes cease? When the system is in equilibrium.

The Statistics of Heat: The Planck-Benford Distribution
Here is a simple formula that allows the calculation of sociological and economic distributions. If you hate formulas, you may skip it.

One might think that equilibrium would mean a state in which all countries have the same amount of wealth. This would indeed be the case, if economies operated according to the statistics and principles of matter—but economies run according to the statistics of heat, where the situation is different. To explain, let us now describe the statistics of heat.

Before examining the distribution of links among nodes, we will ask: what is the definition of equilibrium distribution? What is its meaning, and how can we calculate it?

The equilibrium distribution is defined as the relative probability of each **state** appearing, as compared to the probability of the other states appearing. For example, in a network, a **state** is defined as a node, or several nodes, with a given number of links. The relative probability of finding a node with a specific number of links in the network, as compared to all of the other nodes, is defined as the equilibrium distribution.

A system is in equilibrium when its entropy is at its peak and is therefore stable. The entropy of a system is the logarithm of the number of different combinations of possible states in which the system may exist, multiplied by a constant called the Boltzmann constant. Each combination of system states, which is distinct from other possible states, is called a microstate. Up to a constant, **entropy is the logarithm of the number of microstates.** As a stable system is one in which entropy cannot grow, already being at its highest possible value, we will calculate the equilibrium distribution out of the highest entropy under the constraints of the system. The basic assumption in calculating the equilibrium distribution is that **each microstate is equally likely to occur**.

We will clarify this with an example. Let us say that we have a system with several boxes and several balls, and we want to find the distribution of balls in boxes when in equilibrium. Let us assume that all of the balls are identical, but that each box has a unique location. If we have N balls, each box may contain 0, 1, 2...N balls, and we wish to find the equilibrium distribution of balls in boxes. From a practical standpoint, we calculate the probability of finding a specific number of balls in a randomly selected box.

Surprisingly, even though the distribution of balls in the boxes in equilibrium is unequal, the mathematical assumption underlying its derivation is that **each** microstate is equally likely to occur within a system. This is the reason that an unequal distribution among the system states is counterintuitive—and, some say, unjust as well.

In order to understand the general expression of the equilibrium distribution that will be presented later, let us demonstrate how three identical balls would be distributed among three boxes in equilibrium—that is, we will calculate the likelihood of finding zero, one, two or three balls in any given box.

The number of different microstates is ten. Each microstate is marked as a vector in parentheses—the location in the parentheses designates the box, and the digit designates the number of balls in it. The ten microstates are: (1:1:1) (1:0:2) (1:2:0) (0:1:2) (0:2:1) (2:0:1) (2:1:0) (0:0:3) (0:3:0) (3:0:0). For example, the expression (3:0:0) designates the microstate in which all three balls are located in the leftmost box.

Now, to obtain the relative probability of having three balls in a certain box, we will count how many times three balls appear in a box across all microstates and divide that number by the total number of states. In other words, the probability of finding three balls is three, divided by the total number of states (30—the three boxes multiplied by the ten microstates)—namely 3/30. Similarly, the probability of finding two balls is 6/30, the probability of finding one ball is 9/30 and the probability of finding an empty box is 12/30. We see that the most probable event is that we find an empty box; it is slightly less likely that we will find a box with only one ball, and the smallest probability is that we will find three balls.

We must remember that each of the microstates is exactly as likely to occur as the others, and that in any arrangement of the three balls we will only have one microstate out of the possible ten.

Until this point, we had a simple combinatorial calculation, but when we apply the distribution of balls in boxes to a network, in which we wish to find the distribution of links in nodes, this begs the question: what is the meaning of an empty box? An empty box in a network is actually an unlinked node, which is therefore not part of the network and is not counted among the nodes.[23]

Similarly, when considering heat, an empty state is meaningless, as an undefined number of modes without photons (in other words, empty boxes) exists in the universe. Therefore, in heat statistics, we should calculate relative probabilities while **excluding** the empty boxes. We must remember that at different times, each box may contain any possible number of balls. We therefore only have eighteen boxes with balls now. Three have three balls, six carry two balls, and nine contain a single ball. We see that in equilibrium, the chances of finding one

---

23. Each node may be unlinked at a given moment, but statistical calculations do not relate to the nodes in this level of detail.

ball are 50%, two balls about 33%, and three balls about 17%.[24]

The Maxwell-Boltzmann distribution, which was mentioned earlier in the context of the gas economy, is obtained when one assumes that:

1. The ball-winning chances are the same for each box.
2. The number of balls is much smaller than the number of boxes.

The Planck–Benford distribution is obtained under the following assumptions:

1. The ball-winning chances are the same for each box.
2. The number of balls is much greater than the number of boxes.

Those versed in mathematics and those interested may find the derivation of the Planck-Benford formula in the cited literature (Kafri, 2016). The Planck-Benford distribution, given in the formula below, means that if we distribute a large number of balls into N boxes, the probability of the nth box, $\varphi(n)$, is:

$$\varphi(n){=}\log_{N+1}(1{+}\frac{1}{n})$$

The boxes are numbered in increasing order of the number of balls they contain. $n{=}1$ is the box with the fewest balls and $n{=}N$ is the box with the greatest number of balls.[25]

n, which is the number of balls in the example of the distribution of 3 balls among 3 boxes, may also denote

---

24. These numbers are slightly different from those of Planck-Benford, 50:29:21, because here we examine only 3 balls in 3 boxes, and the general result is for a larger number of balls in 3 boxes.
25. $\log_N X$ is the logarithm base N of X.

a group of several balls taken together, and therefore in physics, n would refer to a rank rather than a number.

In economics, n may be a decile, a percentile, or any other distribution of people into groups of equal size. Since in certain cases the value $\varphi(n)$ in the numerical example that we showed will be the relative probability of finding a box with n balls, in our model, $\varphi(n)$ denotes the distribution of wealth between ranks. We therefore see that the distribution is independent of the absolute wealth of the country. In other words, in equilibrium, **the ratio between the incomes of the deciles in a rich country will be perfectly identical to the ratio between the incomes of the deciles in a poor country**. This is the mathematical expression of the fact that we mentioned earlier: that money is not subject to a conservation law, as the total amount of money in the system is actually zero.

When the number of boxes is large, the probability of a box to be in the nth rank is inversely proportional to n. This means that the greater the number of balls in a certain box, the smaller its relative part out of all of the boxes, in proportion to the number of balls. This law is called Zipf's law. Conversely, in the Maxwell-Boltzmann distribution, the wealth of the country alters the wealth ratio between deciles—that is to say, the ratio between the incomes of the poor and the incomes of the rich changes according to the overall wealth of the country.

The Planck–Benford Law—History and Uses

We will describe how our golden formula was discovered in completely different fields.

The Planck–Benford distribution has different names in different fields, each referring to specific cases of the formula. The law was first discovered in 1881 by astronomer and mathematician Simon Newcomb, from Johns Hopkins University, USA. His work involved numerous calculations, and anyone who frequently made numerical calculations in those days used logarithmic tables, in the same way that we use calculators nowadays. Indeed, until the 1970s, each high-school student would buy logarithmic tables for the arithmetical calculations required in various subjects. Logarithmic tables list common (base 10) logarithms for various numbers. Similarly, there are tables that list the number for each common logarithm—these are called inverse logarithmic tables.

The idea underlying the use of logarithmic tables is that, if one wants to multiply two numbers, one may add their logarithms and look in the tables for the inverse logarithm of the sum—which would be the product of the numbers. In this way, multiplication is replaced by addition, division is replaced by subtraction, finding the power of a number becomes multiplication, and finding the root becomes division. To make it easier to find the numbers in the tables, protruding tabs were attached to the beginning of the tables for various digits, so that any user who was looking for a certain digit—say, five— could find it easily without browsing through every digit on the table. Simon Newcomb noticed during his work that the digit tabs were more worn on the smaller digits.

As one looks for a number in logarithmic tables by its first digit, which is any digit from one to nine (but never zero), he concluded that numbers that began with smaller digits appeared more frequently than those that began with large digits. Surprisingly enough, Newcomb guessed the Planck-Benford law for nine digits ($N=9$), which makes it possible to calculate the relative probability of finding digit n in the beginning of a number. This law is inapplicable to any digit beyond the first, though, because all of the other digits also contain zeroes; thus, the law discovered by Newcomb was named the "first-digit law."

In 1896, Italian nobleman Vilfredo Pareto published his studies on the worldwide distribution of wealth; in them, he found that in many countries, about 80% of all assets were held by about 20% of the population. His work failed to leave a mark at the time, but in 1937, Romanian-born American researcher Joseph M. Juran claimed that the 80:20 rule was a general one. He found that in most cases, about 80% of work in organizations is carried out by about 20% of the employees. In a similar context, we may mention that about 20% of drivers are involved in about 80% of accidents. Juran called this phenomenon the Pareto principle. We will see later that the Pareto principle is another consequence of the Planck-Benford distribution.

In 1901, German scientist Max Planck published an explanation for the frequency distribution of radiation emitted from a "hot" body or, as it is known nowadays, a black body. Despite the title of the topic studied, which sounds like a problem intended for incorrigible nerds (such as physicists, for example), this was a practical problem. Lightbulbs already existed around 1870, and from their invention right up until just a few years ago they were based upon a piece of metal wire with high electrical resistance, heated to a high temperature by an electric current until it glowed. The glowing wire was put in a glass pear full of inert gas, or a vacuum, to prevent it from being burnt by the oxygen in the air. The emission of light from the glowing wire could then be used for lighting. At the end of the nineteenth century, electric lamps were already in widespread use. It is easily understood that lightbulb manufacturers wished to produce the greatest amount of visible light out of the least amount of electricity. Therefore, lightbulb manufacturers supported studies that proposed to explain the distribution of the frequencies emitted by a hot body—which had been measured at the time, but without any theoretical explanation to accompany the measurements.

Planck found a quantitative explanation for the appearance of the radiation spectrum emitted from a given body with a known temperature, regardless of the material it is made of. This work by Planck is regarded as one of the most important works ever carried out in science. The theory of electromagnetic radiation, which encompasses the whole known radiation range—from cosmic rays through X-rays, ultraviolet light, visible light, infrared and microwave, down to radio rays—was known since 1873 thanks to the work of physicist James Clerk Maxwell. Maxwell had found that the speed of light is a constant, and that light is actually a wave. For a given wavelength to exist in a body, the radiation wavelength must be equal to or smaller than its physical dimensions. Therefore, any material body may be described as a collection of boxes, in which the volume of each box is its wavelength to the third power.[26] We may recall that the boxes are not of equal sizes: there are many more boxes for short wavelengths than there are boxes for long ones. In equilibrium, all of the boxes contain (on average) the same amount of radiation energy, characterized by the temperature of the black body.

Planck concluded that the smaller the wavelength, the faster the undulations of the wave and therefore the higher its energy content. He further concluded that the energy of electromagnetic radiation was composed of particles and that the quantity of energy contained in each particle was proportionate to its frequency. Years later, this particle would come to be called a photon.

Planck then solved the following problem: how would the photons be distributed among those boxes of different wavelengths? That part of the black body problem, solved by Planck for the first time, was this: how are identical balls distributed among boxes in maximum entropy—in equilibrium? For the sake of the history of science and as a tribute to Planck's genius, it is worth noting that he was the first to use Boltzmann's entropy formula statistically, and that in our view he was the first to understand the significance of entropy; Boltzmann, who found the correct expression for entropy, had wrongly reckoned that entropy was disorder, when it is actually uncertainty. The formula obtained by Planck was:

$$\varphi(n) = (\frac{1}{\beta})\ln(1+\frac{1}{n})$$

---

26. The assumption is that the box, out of which the photon is emitted, is a cube whose side length is the photon's wavelength.

Where $\varphi(n)$ was the photon's energy as a function of its frequency, n was the number of photons in a mode and $(\frac{1}{\beta})$ was the average energy of a mode that depended upon the temperature of the black body.[27] If we normalize the Planck formula for a finite number of boxes,[28] we obtain the Planck-Benford formula.

This begs the question, why hadn't any scientists noticed the similarity between the two formulas for over a hundred years? This may be in part because it is customary to write the Planck formula in a slightly different way:

$$n = 1/(\exp(\beta\varphi(n)) - 1)$$

Nevertheless, the main reason is the different meaning given to the mathematical symbols. In Planck's writings, $\varphi(n)$ is the photon's energy and n is the number of photons in a given mode; in contrast, in networks, $\varphi(n)$ is the number of links at the nth rank. In probability theory, the meaning is again different: $\varphi(n)$ is the frequency (relative number) of boxes with n balls (detailed proofs may be found in the literature— Kafri, 2016). The Benford law is obtained when we liken the digit n to a box containing n balls, with the constraint that no box may contain more than nine balls. Therefore, the Benford law is a Planck formula, normalized for nine boxes, as Newcomb found in the logarithmic tables.

---

27. The average energy of a mode is kT. k is the Boltzmann constant and T is the temperature.
28. To normalize—to multiply or divide by a constant so that the sum of all probabilities equals one.

In 1935, philologist George Kingsley Zipf researched the frequency of words in long texts and found that the most common word appeared twice as frequently as the second most common, four times more often than the fourth most common, and so on. Zipf, who specialized in Chinese, found that this phenomenon also existed in Chinese texts. We will note that in the Planck-Benford formula, where N is a large number, Zipf's law is precisely obtained. It was later demonstrated that Zipf's law also held for the number of residents in cities, the frequency of earthquakes according to their intensity, and numerous other examples.

In 1937, General Electric engineer Frank Benford showed that in almost any random database he examined, the distribution of digits followed Newcomb's 1881 formula. In other words, if we take a list of share prices, we will find the digit 1 about 6.5 times more often than the digit 9. This also applies to decimal data concerning earthquakes, sizes of cities, or any other random collection of decimal digits. Newcomb's first digit law was then named the Benford Law.

We can see that the consequences of the Planck-Benford law appear in many natural phenomena. If we substitute very small n values in Planck's distribution—much smaller than 1, meaning that the probability of finding a ball in a box is very small and the chances of finding two balls in a box are negligible—the Maxwell-Boltzmann distribution is obtained. The distribution of energy among molecules in gases, which is characteristic to gas economy, and to other distributions in nature that are known as normal, bell-like distributions.

We will now examine the Planck-Benford formula

in our previously presented example, the equilibrium distribution of three balls in three boxes, for which we counted all of the individual combinations. Since N is three, the logarithmic base is $3+1 = 4$ and the Planck-Benford formula will be[29]:

$$\varphi(n)=\log_4(1+\tfrac{1}{n})=\ln(1+\tfrac{1}{n})/\ln 4$$

In other words, we will obtain a 50% probability for $n=1$, about 29% for $n=2$, and about 21% for $n=3$. We can see that our results are similar to those demonstrated in the numerical example for the three balls in three boxes.[30] The probability increases as the number of balls decreases.

Let us now ask: what would happen if we distributed a hundred balls among three boxes? This is an especially interesting question, as in actuality—in many cases—we do have many more balls than boxes. For example, if we have three political parties and one million voters, the individual calculation that we carried out with three balls in three boxes would become impossible due to the vast number of combinations. It bears mention that the probability ratio for the various boxes in the Planck-Benford formula is independent of the number of balls, and therefore, according to the formula we obtained, which describes a state of equilibrium, the ratio between the number of votes won by the various parties is independent of the number of voters. For example, with three parties and one million voters, the ratio between

---

29. $\log_N X=\ln X/\ln N$

30. The analytical Planck-Benford law is accurate for a very large number of balls.

those who vote for each of the three parties would still be 50:29:21.

To demonstrate the usefulness of the Planck-Benford formula, let us study the following figure, which shows the distribution of seats in the Israeli Knesset among the various parties in Israel's 2015 elections. We see a high correlation between the distribution of Knesset seats among the parties and the Planck-Benford distribution. It is worth noting that a similar distribution was found to apply to many elections held in Israel, with the exception of those held during the earliest years of the state.

The Distribution of Knesset seats in the 2015 elections in Israel

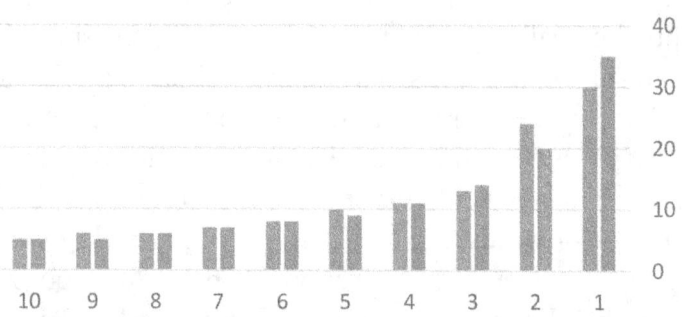

Figure 4: The distribution of Knesset seats among different parties in the 2015 elections in Israel. The right-hand columns are the theoretical results and the left-hand columns are the actual results.

The predicted distribution of Knesset seats in equilibrium according to the Planck-Benford formula was more accurate than any of the 25 polls conducted by various polling companies before the elections. The "drawback" of the formula is that no reference is made to the actual parties. We can predict how many seats will go to the third-largest party, but have no way of knowing which party that will be.

Surprisingly enough, the Planck-Benford formula is used nowadays to detect financial statement fraud and little else. In the investigations departments of the income tax authorities in several states in the US, the distribution of digits in financial statements is examined by a computer. If the distribution is inconsistent with the Planck-Benford law, the statements are referred to tax investigators for closer inspection.

The Planck-Benford formula can be used in bestseller lists (where books are boxes and readers are balls), in the distribution of city populations (the cities are boxes and the residents are balls), and in many other subjects.

Chapter 3

The Statistics of Money

In this chapter, we will see that the distribution of money follows the Planck-Benford formula even though we assumed perfect randomness when deriving it. We will further explain why the constraints imposed by states and people do not affect the distribution.

Constraints and economic inequality

The heat economy, based upon maximum entropy, assumes perfect randomness, where each microstate within the economic network is equally probable. Any constraint imposed upon the economy reduces the number of possible microstates, and as a result, entropy decreases and the economy appears to depart from its optimum state, the state of maximum uncertainty.

We will see, however, that Planck-Bedford's heat statistic surprisingly works, even though a constraint-free economic system does not actually exist. We will assert that the constraints themselves are random by nature, reflecting, for the most part, a war of interests between nodes. Since the distribution of revenue among nodes is independent of the volume of trade, the system will reach an equilibrium distribution even when the constraints cause a general slowing of trade.

The state runs the largest system of constraints through the legal system and through regulation, varying from country to country according to their prevailing philosophies. Furthermore, people also impose constraints upon themselves, having to do with the social network to which they belong, such as the laws of the religion they believe in, or the regulations of the companies in which they work, such as confidentiality and non-competition agreements.

Most laws are driven by the concept of justice and the moral values of a majority of citizens of a given country at a given time. Justice and moral values are subjective concepts that vary widely over time. In the nineteenth century, many people had their daughters marry at the age of 12, whereas nowadays it is illegal, in most Western countries, to marry before the age of 17, while sexual intercourse between an adult and a 12-year-old minor is seen as rape, punishable by imprisonment. In the 1950s, homosexuality was a jailable offense. Nowadays, homosexuals may legally marry and adopt children in many places.

We tend to believe that laws are based upon some logical foundation, but things are not so simple. Here is one example from Judaism: the phrase, "Thou shalt not seethe a kid in its mother's milk," (Shmot 23:19)[31] appears in the book of Exodus. This is a moral edict instructing us to sanctify the continuity of life, even with regard to the life of a beast originally raised by Man to serve as food. Years later, Rabbis expanded this ban into a rigid system of rules, dictating the separation of all foods containing meat from all foods containing milk[32], one of the main rules being that separate dishes and containers

31. Meaning: "do not cook a young goat in its mother's milk".
32. Except for fish and grasshoppers.

must be used for serving and cooking each type of food. In modern legal parlance, this means that where the law is "Thou shalt not seethe a kid in its mother's milk," the binding precedent is that one may not cook any dish containing meat—such as chicken—in a container that was formerly used to cook milk or any other dairy dish. Common law is a basis for prolific economic activity. For instance, in the case of the above example, as well as other religious food-related rules, a 'kosher industry' was created and became a meal ticket for many people. In the United States alone, this market has a turnover of $12.5 billion (this activity seems non-productive for people who do not observe the kosher lifestyle).

European liberalism, whose name derives from "liberty"[33], became a religion that imposes upon us certain rules of behavior. Philosopher John Stuart Mill said that the meaning of liberty is that society may only intervene in the lives of individuals for self-defense—i.e. to prevent harm to others by an individual. In this spirit, one may ask: why is it that in many places where two men or two women may marry, it is forbidden for a man to marry two or three women and vice versa, even where all of the interested parties wish to do so? Among the sanctified values of the "liberal religion" today we may find the sanctity of life, the freedom of speech, academic freedom, and the right to possess property. We will later see that various groups exploit these values to their own benefit, using binding precedents decreed by courts while twisting their meaning to accumulate wealth, in a similar manner to the way the edict, "Thou

---

33. The origin of the name is the Latin word "Libertas", literally meaning "liberty".

shalt not seethe a kid in its mother's milk," is treated in Judaism.

Were it possible to enact laws and set common laws that follow cosmic logic, we would live in a deterministic world, in which each problem has one correct solution. Were such a world to exist, it would be theoretically possible to build a "super-computer" that would find for us the correct decisions that we should make. The conclusion would be that the current state of such a society would allow it to accurately and consistently predict the future.

If we go back to the ball and box statistics, each opinion, as silly as it may seem to us, is a box and the people, who are the balls, are distributed among the opinions according to the Planck-Benford law. Why? Because the numerous laws, which change very frequently, lack cosmic logic and do not alter the distribution, which we have seen to be universal. Nature cannot distinguish between good and evil. Good and evil are determined as the result of a struggle between interest groups, which subdivide to maximize entropy. Therefore the economic system, with all of its constraints, maintains the same statistic of boxes and balls. The cruel statistic is that the number of poor people greatly exceeds the number of the rich.

Anyone would agree that poverty is an evil to be fought. Let us then consider what would happen if every person in the world were rich. In that case, who would perform the low-paying jobs—who would serve food in restaurants? Who would change the diapers of nursing patients? Who would collect garbage in the streets? The poor are an important part of the economic network. When a country does not have a sufficient number of poor people, it must import them from poorer countries, with all of the associated social implications. This happens in many developed countries, which bring people from Africa and Asia to do all the work that their richer citizens do not wish to do.

Could one create an economy without rich people? This idea may be popular. The number of the excessively rich is much smaller than the number of the poor, and therefore most people not only object to poverty, but also to exaggerated wealth.

Well, let us return to our assumption of money being heat. When the tycoons receive big loans from the banks, they usually spend the money on projects that employ workers, to the benefit of the entire population, since the tycoons' employees receive salaries that allow them to pay the food manufacturers and the suppliers of services and goods. Therefore, tycoons play an important role in an economy.

Any liberal democracy has a legislative authority that invents new laws, allegedly to increase justice and raise the quality of life for residents of the country. In many cases, the state expands Mill's principle, according to which society must not interfere with the lives of individuals, except in order to protect others, enacting laws and establishing common laws that are only intended to gather more money and power. In Israel, for example, there is a law against gambling. It is very hard to explain how investing in the stocks of a publicly traded company is any different from gambling in a casino. Moreover, even though the state bans gambling, it operates two monopolistic gambling companies of its own—one named "Mifal HaPayis" (its best-known betting game is the "Lotto"), the other for sports betting (its best known game being the "Toto"). It bears mentioning that the ratio between the prizes distributed to winners and the revenues of both of the gambling companies owned by the State of Israel, in which gambling is prohibited by law, is vastly smaller than the customary ratio in any legal privately-owned casino around the world, including those held by crime organizations. To put things in perspective, Mifal HaPayis distributes about 58% of its revenues as prizes (2015), whereas a casino generally distributes around 95% of its revenues as prizes. The state thus exploits a human failing and creates a predatory monopoly with its gambling operations, while making competition illegal.

Drugs, alcohol, prostitution, and gambling have been part of human behavior since time immemorial. Between 1920 and 1933 it was forbidden to sell alcoholic drinks in the USA because alcohol is an addictive drug. Those who benefited the most from this law, though, were the crime lords. Nowadays, even though alcohol is every bit as addictive as it was in those days, it is a perfectly respectable and even prestigious occupation to own, for example, a whiskey brewery.

Should laws be discarded, where they are inherently illogical? There is no clear-cut answer to this question. When drugs are made illegal, their market contributes to a different type of economic activity. Were it legal to sell drugs, economic activity would be in the hands of farmers, pharmaceutical companies, and pharmacies. Since it is banned, the result is that many burglaries are committed by desperate people, who need money to purchase drugs whose prices are high because it is forbidden to manufacture and sell them. This provides a living for policemen, insurance companies, lawyers, judges, wardens, social workers, workers in drug rehabilitation facilities, and drug dealers. Therefore, most of the people who make a living out of the ban on the drug trade are normative, positive people. In 2010, the direct spending of the USA on the war on drugs was $40 billion, out of which $13.5 billion was spent on fighting marijuana alone. 11,000 people are employed by the Drug Enforcement Administration in the US. And who objects the most to the cancellation of the ban on drug sales? The police itself. What economy-oriented company would agree to have such a bonanza taken out of its hands?

This same phenomenon repeats itself with prostitution and gambling. A person who runs a brothel does not pay taxes and therefore need not worry about the social rights of the prostitutes. One may safely assume that the medical examinations desired in this occupation are conducted rarely, if at all. In the final analysis, the public pays for the ban on prostitution not only with the black money circulated in the industry but also in financing the normative people who benefit from the law: social workers, the police, the local authority employees, sex doctors, etc. It seems that, more than the criminals themselves, the main beneficiaries are elected officials, because any increase in the number of civil servants performing such 'sacred work' is accompanied by corresponding increases in the budgets and the power of those employees.

Imagine the disaster that would befall the state attorney's office, had crime ceased to exist, and conversely, imagine how good it would be for that office if an offense could be attributed to anyone and everyone! As Lavrentiy Beria said long ago, while he was the head of the NKVD in the Soviet Union, "Show me the man and I'll show you the crime."

Here lies the great paradox: economic activity is economic activity. It doesn't matter if it is legal or not, and therefore drugs, prostitution, and gambling are economic industries that provide a living for the population. The conclusion is that the influence of those constraints known as laws and common laws on the economy may alter the identity of those who stand to gain or lose, but not the distribution itself, since—from an economic standpoint—both the pharmacist who sells legal drugs and the attorney who prosecutes drug dealers are boxes in a statistic.

In the free world countries, regimes are found on a spectrum between two diametrically opposed approaches. On one end of the spectrum, there is the libertarian-capitalistic approach, in the spirit of Milton Friedman, who said that liberty reigned supreme and each person was free to do as they wished so long as they did not harm others. The meaning: minimal regulation, minimal government services, and minimal taxation. At the other end, there is the social-democratic approach (its prevalent version nowadays being the welfare state), in which social security and the right to live respectably are supreme values. The meaning: the state will hold monopolies in fields such as railway companies, electricity, water, etc., and will provide its residents with the maximum governmental services possible in return for charging them high taxes. Actually, no existing country totally occupies any end of the spectrum. In the USA, when the Republican Party is in power, it represents the closest approach to libertarianism, professing maximum freedom. France is closer to the social-democratic approach, in which, on one hand, the state provides more services, such as superb public health services, and on the other hand, it prohibits Muslim women from wearing burqas in public places—a ban that would be unthinkable in the USA or in Israel.

As a matter of fact, both "freedom above all else" and "security above all else" are nothing but slogans. By "freedom," we mean that each person is responsible for himself, in the spirit of Mill's words, an idea that implies minimal regulations. When someone is incapable or unwilling to be responsible for himself, for whatever reason, the state will do the bare minimum to help him survive, while preventing him from becoming a criminal or a burden upon society. Many people are willing to give up a part of their freedom for security. Imagine a company whose employees have tenure and handsome salaries. The factory employees sleep well at night because their salaries are guaranteed, unlike the owner of the factory, who does not sleep well at night because he has to make sure that he has money to pay his employees' salaries. The owner seems to have freedom while the employees are secure. In fact, they are all in the same boat, the owner and the employees having the same amount of financial security. Similarly, one cannot separate freedom from security in other contexts as well—even welfare states employ different economic models and include laws that combine both approaches. Let us take motorcycle riders: by law, every motorcyclist must both insure himself and any third party he may hit. His motorcycle must be annually tested in a government-approved institution. As motorcyclists are involved in many more accidents than car drivers, their insurance premiums are higher. This is a consequence of a capitalistic-libertarian approach: the insurance company spends more on the average motorcyclist and therefore charges him a higher premium. There is another law that obligates the motorcyclist to wear a helmet. If caught by a police officer while riding without a helmet, he will pay a fine. This is a socialistic-paternalistic approach. The capitalistic approach, in the spirit of John Mill, would have the rider himself decide whether to wear a helmet and bear the consequences of his decision. When society assumes responsibility for the individual, it dictates how that individual will live. Apart from reducing the rider's freedom, the public also pays for the additional police officers, judges, and attorneys needed to effectively enforce these restrictions.

Another example is smoking. A habitual smoker probably jeopardizes his health and reduces his life expectancy. Even though it is the individual who decides to smoke, the state holds itself responsible for the consequences of that decision and finances the treatment of the diseases caused by smoking. It is not only the smoker himself who pays for the treatment—by law, the general public pays a non-progressive[34] tax for an obligatory national health insurance that every Israeli must purchase. The capitalistic approach would have a smoker pay an additional premium, in respect of the financial risk involved with smoking. Nevertheless,

---

34. In Israel, progressive taxes are only a function of income.

the health tax cost is determined only by the income level of the payer. Therefore, rich people who never smoke bear the treatment costs of the heavy smokers. Many people would say that this is just, but this method does not help eradicate the misguided phenomenon of smoking. Usually, welfare states find a compromise: while caring for the victims of smoking, they lay a tax on cigarettes (and also alcohol, for example). Theoretically, the collected tax funds are meant to partially finance the medical costs incurred by smoking.

How does it come about, then, that social evolution encourages the general human phenomenon of having a multitude of constraints—a phenomenon that expressly contradicts the simplest economic logic? The reason is that sick people, nursing patients, and many other groups also generate economic activity.

Formerly, when economic activities mainly revolved around satisfying basic needs, sick people and the elderly were an economic burden, not sustainable by many societies. Therefore, the Eskimos used to remove the elderly from the community, sending them to die somewhere in the cold; in Sparta, weak children, mentally handicapped, and elderly people used to be killed; in the Fiji Islands, the elderly used to commit suicide, and so forth. In other societies that idolized beauty and strength, such as Rome, the elderly were not appreciated but no such actions as described above were undertaken. Few cultures, such as the Jewish and the Chinese, advocated respect for the elderly. Nowadays, however, the care of nursing patients has become a productive economic activity, similar to the care of other weaker groups such as the chronically ill, the mentally and/or physically disabled, and even the very different field of the care of animals.

Laws and constraints lack cosmic logic due to the complexity of social systems that preclude rational decisions—if decisions were rational, the field of economics would not exist. We still believe that we, humans, strive to behave in a human and rational manner—but what is rational behavior?

In everyday language, we say that rational behavior is an outcome of rational thought, whereas irrational behavior is a result of emotional decisions. The problem is that this definition is only meaningful in a world in which each result has only a single cause; for example, "The vase broke because it fell off of the table." But in the real world, there are many reasons for any given change, and one cannot usually predict results if one focuses on only a single reason. As an example, let us discuss the argument that the Earth is warming because we pollute the air. This is the same "iron-clad logic" that led to the conclusions of the futurists who wrote "The Limits to Growth", on the depletion of resources. Without discussing the problem of why the Earth is warming, let us instead simply state that animals burn oxygen and generate carbon dioxide, while plants consume carbon dioxide and generate oxygen. If we generate more carbon dioxide by our actions, more algae will populate the sea, and more forests will grow in the land, which will, in turn, generate oxygen out of the carbon dioxide. In the last century, Earth´s magnetic field was weakened by six percent. This change affects the ionosphere that protects us from cosmic rays. Nobody knows how this affects global warming. On September 11th, 2001, an unbelievable terrorist incident occurred:

two passenger planes crashed into the twin towers in the USA, making them collapse. A third airplane hit the Pentagon. As a result, no plane flew in the US skies for two days. Let us guess what should have happened to the average temperatures as a result: warming, cooling, or nothing? What was observed was warming, by roughly 2°C. It is easier to explain things after they occur, and the explanation given was that the interruption of flights reduced cloudiness, which led to more radiation entering Earth and therefore heating. In this case, does the cloudiness generated by the gases formed out of burnt fuels and various other troublesome materials lead to cooling or to heating? Would you have bet your money on this answer before you knew it?

Another argument often raised in the debate on global warming is the issue of the melting icebergs in the North Pole—we see this very often in scary films and TV productions. Few people know that at the same time, the area of the icebergs in Antarctica actually grows.[35] Additionally, icebergs do not follow a uniform pattern as they melt in the North Pole. One should remember that the generation of icebergs also depends upon the amount of precipitation and on many other factors; the facts we present here do not support or refute the theory of global warming. The logical conclusion is that it is impossible to know whether the Earth is warming, and if it is warming, what the cause is.

Another example of a simplistic approach to a complex problem is the apartment market in Israel: following a recession in the real estate market spanning more than a

35. It bears mention that apparently icebergs melt at the North Pole at a greater rate than they grow at the South Pole.

decade, a shortage of apartments was created. In 2007, after the capital market entered the subprime crisis that we will discuss later, the yield on capital decreased dramatically, making residential real estate a hot investment, followed by a steep increase in price. This is a natural, cyclical phenomenon, one which is usually corrected in a normal market by accelerated construction of apartments by the private sector. However, rather than let the market adjust itself, the government decided to intervene. In 2014, the minister of finance at the time proposed to grant a VAT exemption to those who purchased their first apartment from a contractor. The proposal sounded wonderful at the time; it represented an 18% discount on apartments for first-time apartment buyers (even though the benefit was offered to those well-to-do people who could afford to purchase apartments, and therefore generated great opposition). But what actually happened? By the time the proposed act cleared all of the committees and votes, contractors were delaying construction and buyers were putting their will to buy on hold, as everyone was waiting for the coveted discount. Finally, the tenure of the government ended prematurely for other reasons altogether, the act fell through, and the idea—though basically good—exacerbated the real estate crisis because the construction rate slowed in response to hopes for the discount. Then another minister of finance came, in whose view the shortage of apartments, like the other woes of the world, was caused by what he called "the apartment collectors"—people who had a greater number of apartments than he considered necessary for residential purposes. In his opinion, the correct solution was to tax them (it is never wrong to tax the rich, especially during an electoral campaign, as there are more poor voters than rich ones). However, as the "apartment collectors" rented them out to those who could not afford to buy an apartment, it is clear that not even a single apartment would be created thanks to the new tax. Furthermore, the tax will drive the "apartment collectors" out of the market, thereby reducing the demand for apartments and the contractors' motivation to build them. The correct solution would have been to maintain stability (or in other words, to do nothing) and let the market take care of itself.

The serious definition of the term "rationality"[36] comes from a branch of mathematics called game theory. In game theory, it is assumed that in a multiple-player game, all players will always act to maximize their own benefits. In other words, each player will consider all of the possible actions and choose the sequence of actions that would generate the best result for him. This is rational behavior. Kahneman[37] and Tversky showed, following that same iron-clad logic, that people do **not** behave rationally. One of the many examples they gave (or that were attributed to them) was the discussion of what people would do, having purchased a ticket to a show for a hundred dollars and then lost it, compared to those who discovered that they had lost a hundred dollars as they were approaching the box office to buy a ticket. According to polls conducted by Kahneman

---

36. In philosophy, unlike economics, rationality is defined as preferring inference over empiricism. We are not referring to this definition.

37. Daniel Kahneman won the Nobel Prize for Economic Sciences in 2002.

and Tversky, most of those who purchased a ticket for the show and later lost it would not buy a new ticket and would give up the show. Conversely, most of those who lost a hundred dollars before buying the ticket did not give up the show. A hundred dollars being a hundred dollars, according to Kahneman and Tversky that person should rationally behave in the same way in both cases. Another example of our own: a man dines at a restaurant, receives superb service and gratefully leaves a tip of NIS 30 instead of the normal NIS 15 that the billed amount would warrant. Later, he takes a walk and becomes thirsty. He enters a kiosk, asks for a bottle of water, and the vendor asks him for NIS 15 instead of NIS 8, the price he normally pays. The diner refuses to pay and continues to walk until he finds water for a non-exorbitant price. Did our walker make a rational decision? Well, his decision depended upon several factors. For example, how thirsty was he? How much time, in his estimate, would he have to wait until he found a place where water was reasonably priced? How much is he generally irritated by people who charge exorbitant prices? And so on. That same person may make different decisions concerning the same situations, influenced by his mood and other circumstances that change over time. In Kahneman and Tversky's example, the decision whether to purchase an extra ticket or to give up the show is one that depends on the circumstances and that therefore has no rational answer. For example, if this is a one-time show by an artist that you are a fan of, you will think twice before giving up the show. Conversely, had you bought the ticket simply because you were bored and were looking for some entertainment, your decision would depend upon the amount of hassle involved in the act of buying a new ticket: how long would you have to stand in line? How significant was the price of the ticket for you? I will add a personal story in this context: during a vacation in Paris, my wife and I had to appear at the airport at 17:00. We had two simple options: one, taking a direct bus for about twenty euro, or two, taking a cab for about fifty euro. To take the bus, we had to walk for about ten minutes from the hotel to the bus stop. Since we had three suitcases and two backpacks, and since thirty euro, the difference in cost between those options, was an amount that we could afford, we decided without hesitation to make the rational decision and reserve a cab for 16:00—but in the end, we travelled by bus. Why? Because even though we had a late-departure arrangement with the hotel, we were thrown out at 13:00. As it was a lovely, pleasant day and the suitcases rolled well on the sidewalk, we changed our decision and walked to the bus stop. The bus was not crowded, we saved a bit of money, and we had fun. The rational decision was still a cab trip that would allow us to arrive comfortably—because it might have rained, because of the tiring walk to the bus stop, and because the bus might have been crowded with passengers. What finally occurred was that a combination of circumstances turned the irrational decision into a correct one. Therefore, rationality is an important theme in well-defined problems such as one encounters in exact sciences and mathematics, but not in economics.

In Israel, there are two Nobel Prize Laureates who are experts in rationality: Daniel Kahneman, who was mentioned in the previous paragraph, and Robert John Aumann, who won the Nobel Prize in Economic Sciences (2005). Kahneman is a leftist politically, while Aumann belongs to the right wing. Is it not a paradox, that two renowned rationalists hold opposing views? Is it possible to explain this paradox by stating that rationality does not exist? The above discussion goes to show that behind justice, law, binding precedents, collective freedom vs. personal freedom, and other similar concepts, there stands an all-guiding law, a natural law—the second law of thermodynamics; a law stating that the uncertainty in any system tends to grow to a maximum, and in other words, whatever the state of the system, it will strive to reach equilibrium, where the distribution is expressed by Planck-Benford's law.

A short summary up to this point: In any country equipped with technology that enables a tiny part of the population to satisfy the basic needs of the entire population, any human activity has an economic meaning and each person seeks to divert the flow of money towards himself. Laws, welfare and crime are all part of this activity. As in the cynical maxim that more people live off of cancer than die of it, one may also say that more people live off of the laws and their enforcement than off of crime. As we said, law enforcement, public health, gambling, and prostitution—most of these are legitimate economic activities. If this is the case and everything is indeed random, let us check how the Planck-Benford distribution compares to the real economy.

The Gini Index
Is inequality necessary, and is there an optimum percentage of poor people?

It is widely agreed nowadays that wealth should not be equally distributed. This was not always the case: a hundred years ago, countries such as the USSR and China believed that all men should share wealth equally. Other countries had also adopted the communistic approach and offered, for example, a superb education and health services to all for free. Nevertheless, the gap between the **welfare** of the citizenry of those countries and the welfare of those residing in countries that were running a free economy was so great that the USSR (which has since become Russia again) and China became capitalistic countries without any additional bloody revolutions.

An outstanding example of the failure of the egalitarian society and one that deserves special mention is the kibbutz. The kibbutz is an Israeli invention that was founded upon communistic ideology, according to Karl Marx's slogan, "From each according to his ability, to each according to his need." In other words, a total separation between one's contribution to society and the material recompense one receives back. The kibbutz is special in that it was an elitist association of people who had a similar cultural background, held similar views (communistic and socialistic), and championed values of equality, fraternity, and peace. The kibbutz was a noble human effort, worthy of admiration, certainly in comparison with communism, which was forcefully imposed upon whole countries and required a dictatorship to exist. Despite the superb human quality of the members of the kibbutzim, the kibbutzim failed financially when the labor party, a considerable number of whose leaders came from the kibbutz movement, fell out of power in favor of more right-wing parties. The new forces in power took some of the financial benefits given to the Kibbutzim by the labor party. The kibbutzim, like the communistic societies, are egalitarian only in an economic sense. Social connections, which are akin to money in any network, are distributed just as unequally as money. In a kibbutz, some people are more connected and some less so, and therefore some people are "worth" more and some people are "worth" less.

In 1912, Italian statistician and demographist Corrado Gini suggested an index for economic inequality. As the Gini Index is not mathematically complicated, let us describe it using common language. Let us divide the population of a given country into ten groups of equal size.[38] We will mention that the statistics measured by the Gini Index do not concern economic entities— only people. Call the group of those with the highest revenues the "top decile." The average revenues of the top decile are, accordingly, the "top decile revenue." We will obviously find the richest person in the country, generally a billionaire, in the top decile. We may also find your neighbor in the residential building you live in, and maybe even you, a senior employee who might be fired tomorrow.

The situation is similar in the bottom decile: we find there the portion of the population that is on the dole, many of whom make a living 'on the side' and do not

---

38. One can do this with any arbitrary number of groups, provided that each group contains the same number of people.

report that fact to the tax authorities, alongside many unfortunate people, usually older people, who have to live on national insurance benefits alone. The second to ninth deciles are a more precise reflection of reality, as the revenues of people in those deciles are relatively similar. Had the average revenues of all deciles been equal, each decile would receive one-tenth of the national revenue. Therefore, the two bottom deciles would receive 20% of the overall income, the five bottom deciles would receive 50% and so on such that the aggregate revenue of the deciles from the poorest to the richest would be a straight rising line, starting at one-tenth and finishing at one. The bottom deciles actually earn less than the top deciles, though. Obviously, in all cases, the total amount of revenues earned by all deciles is 100%.

The Gini Index is the sum of all of the differentials between an equal distribution and the actual distribution. For example, if the bottom (tenth) decile earns 5% of the total income, but would earn 10% in an equal distribution, we should write '5%' next to the bottom decile. If the two bottom deciles earn 11% of the total income and would earn 20% in an equal distribution, we will write '9%' next to the number twenty. If 50% of the population earn 20% of the revenues, we will write 30% next to the number fifty. As 100% of the population always earn 100% of the revenues, the digit 0 will always appear next to the 100% mark. **The sum of all of the differentials that we wrote down, reflecting the differential between reality and equality, is the Gini Index.** If the distribution is equal, the Gini Index will obviously be zero. If the top decile holds all of the money the result will not be one or 100 as one may read in many places, but eighty-two percent.[39] In this respect, the Gini Index is not a perfect mathematical creation, but that does not matter. Since the end of World War II and to this day (2017), the Gini Index varies between 20% and 60%.

The Gini Index has many drawbacks. A quick study of the Wikipedia entry on the Gini Index not only shows how countries with different distributions of revenues may show the same Gini Index, but that studies exist which show how inequality increases in the world as the years go by. This increase completely contradicts everything we see with our own eyes (in economics, statistics are always political and never objective).

This begs the question: what is wealth? Surprisingly enough, many completely anonymous people, who lead modest lives as ordinary citizens, are tens of times richer than celebrities, whose lifestyles are ostentatious. Many of the latter leave a mountain of debts behind them after they die, according to the "economic genre" that suggests that "correct financial planning" is when the check given to the undertaker bounce.

Our economic theory describes money as heat, not energy, and therefore non-liquid assets are not counted (real property, jewelry, art objects, etc.). Many economists define a "millionaire" by only considering liquid assets. For example, if you live in a house worth five million dollars, and you have an art collection worth a million dollars, and nine hundred thousand dollars in the bank in securities, you are not considered a millionaire. Your house and your wonderful collection are not a part of the

---

39. $\frac{10+20+...+90}{10+20+...+90+100}$ =0.82

economy.[40] Conversely, if you sell your house and make use of your wealth, your importance to the economy will grow, because many will come to help you "break free" from your cash. When someone comes to the capital market with a bundle of cash, nobody cares where it came from or who it belongs to; they want to know where the money is going to be routed to and whether any of it will be going to them personally. This is the reason why many people go to politics; politicians route the public cashflow and are twice as respected as a person who works with his own wealth. Unlike a private individual, who cares for his money for the benefit of his progeny, politicians are as generous and merciful as possible, using money that is not theirs but the taxpayers'. The generosity of politicians in routing tax funds comes out of a sense of self-righteousness, intended to flatter the voting public, most of whom have modest revenues. The generosity practiced by politicians is only second to that practiced by journalists, whose generosity manifests in mere words and who scurry to aid any poor and needy person and suggest the generous scattering of monies earned by others.

A somewhat amusing example of self-righteousness may be seen in a recent case, in which a Holocaust survivor was robbed. The journalists came out with sensational headlines, "Holocaust Survivor Robbed!" Did they expect the robbers to inquire, before committing robbery, whether the victim was a Holocaust survivor? This is simply sanctimonious conduct, for which such journalists are handsomely paid. I am always full of generosity and endless love for my fellow human beings when I see the video clip of John

---

40. This matches Kahneman and Tversky's theories, according to which people see changes in their cash flow as more important than the value of their assets.

Lennon playing on his grand piano, in an enchanted castle at the heart of a pastoral garden, that wonderful song *Imagine*, among whose inspirational lyrics one finds the line "imagine no possessions." I cannot help but wonder how wonderful John Lennon's world might have been, had the servants who cared for the castle and the striking garden worked for free and demanded no salaries. In John Lennon's case this was not only about speaking idle words but also about a (wonderful) melody that he had composed, which brought him sevenfold royalties. Sanctimonious journalists influence sanctimonious politicians, who influence the economy. Is this inherent bias in favor of the poor majority taking the economy out of equilibrium?

Before calculating the Gini Index of the Planck-Benford distribution, we must remember that we are about to apply a pure statistic theory in physics to the realm of sociology, into which we humans introduce such concepts as justice and injustice, good and evil, right and wrong, rational and emotional, enlightened and primitive. The findings before us have nothing to do with moral or religious values of any kind whatsoever. The only tool we use is statistics, and some of you may say: so what if we get this distribution through the use of statistics? We, human beings, are better! Or as Barack Obama, the former president of the United States, put it, "We are not like that." We will discuss that matter later on.

The following table shows us the equilibrium distribution of social connections or money among the deciles, according to the Planck-Benford formula. On the left, we have the decile number, and at the right is that decile's pro-rated share of wealth. The top decile is marked 1 and the bottom decile is 10.

Decile	Relative wealth
1	0.289
2	0.169
3	0.120
4	0.093
5	0.076
6	0.064
7	0.056
8	0.049
9	0.044
10	0.040

Table 1: The equilibrium distribution of links or money among deciles, according to the Planck-Benford law.

The following figure shows the Gini Index calculated for that distribution.

Gini Index

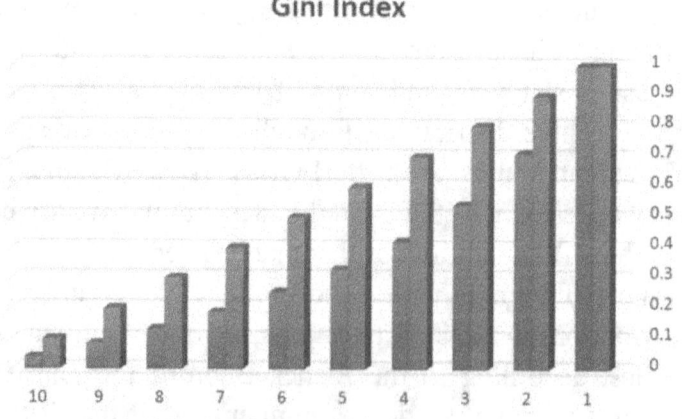

Figure 5: The Gini Index calculated for the distribution of wealth among deciles.

The right-hand columns indicate the aggregate revenue of the deciles given absolute equality, e.g. the sixth column showing one-half of the revenues, or the revenues of the tenth, ninth, eighth, seventh, and sixth deciles. As all deciles have equal revenues, their aggregate amount progresses linearly. On the other hand, the left-hand columns reflect the actual distribution and are therefore smaller than their right-hand counterparts. The greater the inequality, the greater the height differential. The sum of the height differentials is the Gini Index.

As we noted, the differential is always equal to zero in column 1, as the aggregate pro-rated revenue is always equal to one. The Gini Index in percentage points, calculated by the distribution of revenues from table 1 (the Planck-Benford formula) is 32.7%, whereas the average Gini Index for 2015, which was published as these pages were being written (2017) for the thirty-five OECD countries, was 32%. This result demonstrates that the economies of the developed countries are in a state that greatly resembles the equilibrium distribution. When the Gini Index greatly departs from the Planck-Benford formula, this means that the government is running an unbalanced economic policy.

The formula for the Gini Index may be used to express the inequality of any type of distribution, not necessarily only that of wealth. For instance, one may calculate the Gini Index for the distribution of voters among parties in elections: in the elections held in Israel in 2015, ten parties were elected to the Israeli Knesset. The distribution of Knesset seats is shown in figure 4 on page 68. The inequality that we calculated—of the relative number of voters among parties according to the

Gini Index—was 32.4%, a result that greatly resembles the calculated results for economic and thermodynamic inequality. Each point in the Gini Index represents one percent. As we mentioned, the error margin, when performing measurements in the field of economics, is much greater than a single percentage point. We tend to think that in physics, measurements are accurate and precise, but even in physics they are not. The main difference between physics, as a representative of the exact sciences, and economics, as a representative of the sociological sciences, is that in physics, the scientists generally know which quantity they are measuring, and therefore know how to estimate their margin of error.

The amazing thing about the precision level of a Gini Index prediction, when the Planck-Benford distribution is employed, is that when we derived its formula, **the basic assumption was that the probability of each box holding a ball was exactly the same as that of every other box.** I doubt whether anyone would be willing to accept this assumption for human beings. Obviously, the physicist Albert Einstein was not an ordinary box, and neither was author F. Scott Fitzgerald. But let us remember that the second law does not say that the boxes are equal, but that each box has **an equal opportunity**. Some people see opportunity where others not only fail to see it, but even treat it as an obstacle. This is seen in the story of the owner of a shoe factory who once sent two salesmen to Africa. One sent a message to the owner, saying, "It's a catastrophe, nobody wears shoes in Africa!" while the other wrote, "Bonanza, no shoes here, huge potential market!" Which salesman was right? The first commandment of the "liberal religion," stating that each person deserves an equal opportunity, seems patently absurd when one looks at the progeny of the English royal family, compared to an infant born to a tribe in the Amazon. Even a much more talented writer than the author would not manage to convince anybody that their chances of success are equal. If we examine the one hundred richest people on Earth at the time of this writing—2017— all of whom are billionaires[41], we will discover that 73% of them earned their own wealth and only 27% inherited it. Moreover, 36% have no higher education. These data show that there is obviously a real chance for a child born in a family in the Amazon to be richer than some English count or duke in fifty years' time. Who would believe that Alejandro Toledo, who was born into a large, poor family in the Amazon, would later become, in 2001, the president of Peru? But the stronger argument in favor of the heat economy is not the vast number of poor people; it is rather the absurd wealth in the hands of very few people. Why, for heaven's sake, does a person need one billion dollars? What can you do with one billion dollars that you cannot do with fifty million? One billion dollars with an annual yield of 1%—an awful investment by any standard—will still generate about one million dollars per month without ever touching the principal. There is no reasonable way to spend such an amount. In centuries past, the British nobility used to waste its money by building huge palaces that required large service crews and huge amounts of money for maintenance, together with ornamental objects and art, forests, and personal

---

41. About 2,000 people on Earth are worth more than one billion dollars.

servants who helped them get dressed in the morning. We assume that nowadays any normal, healthy person prefers to dress up by himself, regardless of his income. About thirty years ago I used to dictate my letters to my secretary; nowadays, almost nobody dictates letters anymore—people type their own letters because it's simpler, faster, and more comfortable. So why do we run after money when we have enough of it? There are many reasons for that. The thermodynamic reason is that money increases the possibilities available to us (the entropy), and therefore, according to the second law, we want more and more of it. The reason people use to explain away their inherent greed is that great wealth gives them financial security for generations—to them, to their progeny, and to their descendants. Even Warren Buffett and Bill Gates, who both decided to donate their immense wealth to Mankind after their demise, will still leave something for their children.

Let us make an enlightening calculation. Suppose that you put one dollar in the bank for two hundred years, with compound interest of about 6% per year without any linkage. How much money will your descendants have at the end of that period? The answer is 231,656 dollars. In four hundred years this amount will turn into 13.25 billion dollars. Not bad, even considering the likelihood of a stratospheric increase in the cost of living over that long of a period. The original question behind this example is, "Which investment was better four hundred years ago—to buy Leonardo da Vinci's Mona Lisa, already a hundred-year-old and very valuable painting, for a price equivalent to one of today's dollars, or to deposit that dollar in the bank with a non-linked compound interest of 6%?"[42] The correct answer is to deposit the dollar in the bank. If you happen to talk with investment advisers for companies that specialize in investing your savings, and you ask them whether there was a reasonable chance for them to obtain a 6% yield for you (not to mention their own 1% commission), you will make them smile with disdain for even doubting it. Indeed, such yields have been in existence for quite some time and over lengthy periods. But we need to remember that this type of yield is not sustainable indefinitely, only as a temporary fluctuation—a kind of bubble that inflates and soon bursts. I offer a winning tip to the readers of this book on predicting an approaching financial crash (at no extra charge!): the crash will always arrive within thirty days of me investing in a given market. It is lucky for Mankind and the economy that I do not have enough money for my investments to be capable of destroying the whole worldwide economy! By their nature, economic systems cannot remain stable in the long term, and billions will disappear in a crash. The vast amounts earned by people, above and beyond any human need, and in many cases even to their own detriment (the loss of privacy, the need for bodyguards, the negative mental effects suffered by relatives, inheritance disputes, etc.), are convincing proof of the Planck-Benford statistics. But there is an even more convincing proof! How can one lose such enormous amounts of money? It is hard to believe that people with such imaginary amounts of money could lose their entire wealth, even entering bankruptcy procedures—sometimes jail. These are

---

42. For the painstaking reader, let us mention that four hundred years ago the dollar did not yet exist.

people who apparently have much more money than they can ever spend, and yet who somehow seem to lose it all. There is a story about Donald Trump, who once asked his girlfriend during a difficult period, as they were walking along Fifth Avenue in New York, "What is the difference between the beggar on the corner of the street and me?" He immediately gave the following answer: "The beggar is worth a hundred million dollars more than I am."[43]

Billionaires make these enormous amounts of money through the varying value of assets, and lose them for the same reasons. When a tycoon, who made his money through crazy business gambling, loses billions (let us remember that the liquid money is registered as debt, backed by the tycoon's assets) and cannot repay his debts to the bank, it is better for the bank to receive a part of the loan granted to the tycoon from public funds than to receive nothing and have to fight with the other creditors over the apartment, the car, and the jewelry of the formerly rich, now bankrupted individual. Assuming that the tycoon lost his money on a failed venture, his debt generated economic activity in the market, and therefore the public was not negatively affected. The main casualty was the bank, for whom this was the lesser evil. In the nineteenth century, the richest family in Europe was the Rothschilds—reputable, prudent, and trustworthy bankers, who are nowadays worth about four billion dollars. Several Israelis are richer than they are. The Rothschilds did not embark upon wild investments— quite the contrary, their investments are an emblem of conservatism, but it is still unclear whether their capital

---

43. According to the (probably imaginary story), this was the amount of Trump's debts at the time.

will hold good for four hundred years, like the dollar from our story. Decisions made by people, which can never be known before hand to be conclusively good or bad, are the invisible hand that spins the roulette wheel of life. Compare this to the belief that rational thought exists and that the world is deterministic, or that the uncertainty in it is only a result of lack of knowledge and can, therefore, be resolved by research and study. As an aside, we will note that Einstein believed in determinism, and that this was his great error. His aphorism, "God does not play dice," is quoted as the main criticism against him. This fundamental error by one of the greatest minds of all time should serve as a warning sign for us: no one is impervious to errors, regardless of how much of a genius he is.

Poverty and Wealth
The ratio between the rich and the poor, and why they are both necessary

Ben Zoma's Mishnaic maxim: "Who is rich? One who is satisfied with his lot,"[44] refers more to happiness than to wealth. There is no clear correlation between wealth and happiness, but as scientists have not created or identified a "happiness metric" to date, though one may well assume that a certain quantitative connection does exist between the level of certain neurotransmitters in the brain and the level of happiness, one may assume that the distribution of happiness among people would be bell-like and not Planck-Benford's long tail distribution. Had happiness

---

44. Mishna, Pirke Avot 4:1

been distributed in the same way as revenues are, the multitudes of miserable people would probably rise up and exterminate the few happy souls. Wealth, on the other hand, has much more accurate definitions, which can be found in the bank account. In defining poverty there are two approaches: one absolute, and the other relative.

The American approach is the absolute one, which reckons that in order to lead a respectable life, everyone needs a certain amount of money for basic expenses—for housing, food, electricity, water, education, and so forth. According to this approach, each person, being a citizen of the United States, is entitled to a basic basket of goods. The American approach is basically logical. American society is rich, being the most important producer of food, technology, and culture on Earth. The wealth of the United States allows it to grant a basic basket of goods to the poor. In addition, as was amply explained in the beginning of this chapter, helping the poor is an economic activity in every way, and creates jobs in various sectors of the American economy. A further gain for the public is that poor people, whose basic needs are provided for by the state, do not break into other people's homes due to hunger—a common scene in nineteenth-century Europe. In some cases, we tend to approve of government expenses incurred in helping the weaker elements of society out of a selfish motivation. Take the care of nursing and old patients for example—most of the public know that one day they will also be old, themselves a nursing or terminal patient, and that it is, therefore, advisable to provide the best care for that population that they might also be joining one day. Things are similar with the care of the developmentally disabled population. The distribution governing the births of developmentally disabled children is uniform and independent of the economic status of the family—anyone might have such children. A further element that economically justifies the basic support of the poor is the potential value of their gene pool, which generates renowned individuals in almost every field, much in the spirit of the maxim from the Babylonian Talmud: "Be heedful the children of the poor, for from them Torah goeth forth."[45]

Unlike the United States, in most of the world poverty is defined using the relative approach. It seems that the reason for this is the belief held by the "liberal religion", that inequality is the source of all evil. Unlike the absolute American definition, according to which government support varies from time to time according to changes in the cost of living or with varying needs, the European definition is relative, and defines a person as poor if his income is not more than one-half of the median income. The median income is the income level that is lower than the earnings of half of the populace and higher than the earnings of the other half. Those who earn the median income lie between the fifth and sixth deciles. If we look at table 1 again, we will see that those who earn a median income earn about 7% of the general income, compared to the tenth decile, which earns about 4% of the general income. Since economic metrics are not so accurate as to justify a painstaking calculation, our prediction will be that 9% of the residents of a country would be poor, based on the Planck-Benford distribution. We will mention that

45. "Beware the sons of the poor, out of whom Torah will come", Vows Tractate 81:1

this is exactly the percentage of poor people found in the OECD nations.

There is a basic problem with the definition of **relative poverty**: poverty is not necessarily related to real scarcity. This problem has two aspects: firstly, a person who lacks nothing and who may even be happy with his life may still be defined as poor; secondly, it is possible that a person who lacks the basic conditions for survival would not yet be defined as poor. If we think that these are only the hypothetical assumptions of a theoretician, let us observe the communistic countries for an example. We will find that in certain years, even certain cross-sections of the upper-middle class were suffering from nutritional insecurity.[46] A friend of mine, a senior physicist who emigrated from Poland to the United States in the 1980s, once said to me while I was staying there on a sabbatical: "Life has no meaning in the United States." When I asked him what he meant, he replied: "In Poland, when I obtained two pounds of butter I was happy; when I found a fish in a shop the entire family was happy. Here you enter a supermarket and they have it all." Another example is the lobster, one of the most luxurious dishes in any elite restaurant, so much so that the 15th of June is celebrated as Lobster Day in honor of the well-known delicacy. Compare this to the colonial period, when lobsters were regarded as food for servants, prisoners, and slaves. Employees went so far as to demand an article in certain employment contracts stating that Lobster may

---

46. Nutritional security is defined as a person's ability to provide himself and his family with healthy, nutritious food that contains all of the main food groups in sufficient and satisfactory quantity and quality.

not be served to them more than twice a week. The same dish that used to mark one as poor and wretched is now perceived as a symbol of wealth.

When I was a child in the 1950s and the cucumber season started, my mother used to slice them into our salad. The wonderful smell of those cucumbers is still vivid in my mind. Now we call these smells and tastes "the flavors of yesteryear." Why do we pine for them so? If the taste of cucumbers has changed at all since that time, it was only for the better. The thing that changed was that nowadays we can have a green cucumber throughout the whole year, its price is low and very affordable, and so we no longer savor its gentle smell. To quote from Ecclesiastes (9:16): "The poor man's wisdom is despised, and his words are no longer heeded."

A virus attacked the tomato bushes in Israel in the spring of 2016 and the price of tomatoes rose sharply. I do not want my words to be seen as any disrespect to tomatoes, but even nutrition experts would say that one can live without eating tomatoes every day. Our generation, which remembers the way cucumbers used to smell, also remembers that tomatoes used to be eaten in their season, as greenhouses did not exist at the time.

When the price of a certain commodity rises, it is consumed less. This is the way of the market. The Strauss family, which came to Israel from Germany in 1936 when the Nazis rose to power, started a dairy farm to make a living. The farm eked out a bare existence in its early days, but developed throughout the years, underwent changes, and is nowadays an international food conglomerate that specializes in dairy products and other delicacies and employs about ten thousand employees. Following the sharp increase in the price of tomatoes, due to the virus that attacked the bushes, a female member of the Strauss family "dared" to say "If it is expensive, they shouldn't eat it. Who said that one should eat tomatoes all year long?"[47] Indeed, tomatoes were a seasonal fruit in her day, and the world didn't come to an end. She recommended that people wait for the price to go down in such cases, and then eat tomatoes again. I cannot fully recall how the minister of finance answered this statement, but I remember well the disgusted expression he wore when he said that a woman who can buy almost everything her heart desires should not dare disrespect the tomato of those who have nothing. His fiery speech was broadcast over and over again—how does a woman who can buy almost everything her heart desires dare disrespect the tomato of those who have nothing? The Strauss factories generate a living for over ten thousand people and pay about one billion NIS in salaries each year, net economic activity regardless of the profitability of the factory, which is only important to the few shareholders.

The person in charge of the Israeli economy tries to please his voters more than their employers—and to promise them cheap tomatoes. The minister of finance behaves in a human manner: he takes care of his wallet and his seat. Anyone not involved in growing tomatoes loves the landlord who promises him cheap tomatoes. The problem is, how do you supply tomatoes cheaply when the tomato bushes are attacked by a virus? Even if a genius inventor were to invent a wonder tomato that could be distributed for free, soon enough nobody would

---

47. Calcalist, 14.4.2016

want it, just as lobsters were not wanted in the past. Had our minister of finance lived in nineteenth-century Boston, he would probably decry the fact that the poor are forced to eat disgusting vermin from the ocean, barely worthy of serving as food for prisoners.

It is essential to have an unequal distribution of money. If everyone is well off, nobody will want to engage in difficult jobs that require physical effort. In the western countries, there is nowadays a severe shortage of local workers who are willing to perform simple jobs, even in professions that were considered respectable until a few decades ago, such as masons, plasterers, and floorers, only because these professions require physical effort. Immigrant workers from foreign countries now perform these and similar jobs. The reader is reminded that each citizen in the western countries is guaranteed a minimum-level income if he does not work and meets certain requirements. Therefore, if someone finds work for "black" money, which he may receive without an invoice or a pay stub, he will receive the minimum-level income from the National Security Institute in addition to his wage. This is a common situation that makes the official level of poverty higher than the actual level. The proof is that workers have to be brought in from faraway countries to care for the elderly and do construction and cleaning work.

It would be shortsighted to think that this is only a problem in economics. This problem exists in all fields, because the Planck-Benford distribution is universal. For example, nobody would dare to say that education should not be given to all. Indeed, almost all countries provide their residents with subsidized education, and everyone is happy to speak of their academic degrees. The percentage of degree holders keeps increasing and lo and behold— the job security of these new scholars does not increase, because multiple academic degrees do not guarantee a living. If everyone were to become professors, who would operate the manufacturing machines of industry? Who would become a plumber, a profession that affords a handsome living to those who master it?

Poverty is a rather vague concept. According to the Gini Index, the level of inequality in Israel is above the OECD average. One may, therefore, suppose that the poor would number more than 9% of the population. According to an article by Amatzia Samkai in Ma'ariv, published on December 12th, 2016, only 6.4% of those interviewed in a survey conducted by the "Latet"[48] association (an association founded to help poor people) see themselves as poor. We understand, therefore, that poverty is a subjective state, one not related to scarcity— at least in the scientific sense. The coalition parties would generally try to present a lower-than-actual poverty rate, while the opposition parties would try to present a higher-than-actual one. The opposition seeks to convince the voters that the corrupt coalition government is the source of our troubles, while the coalition wishes to show that it runs the economy in the best possible way.

The "Latet" association subsists on donations, feeds the poor, and pays handsome salaries to those engaged in their holy work. According to Parkinson's Law, "Latet" and other similar associations would like, consciously or otherwise, to increase the percentage of the poor, and

---

48. Meaning in Hebrew: "To give"

thereby, in all likelihood, the number of their employees. Following that logic, a government committee will destroy itself if it solves the problem it was created to solve. According to the "Latet" association, the poverty rate in Israel is 29% (2016). Why? The association defines "poverty" by certain criteria, and if the person surveyed does not meet some of them he is still defined as poor. For example, if you have no academic education, you have met one poverty criteria. One criterion is not enough for you to be classified as poor; several more are required, such as more than two persons per room and an inability to purchase complementary medical insurance. Remember, though, that 36% of the hundred richest people on Earth lack higher education. If we further examine the severe implications of a society entirely composed of university graduates, most of whom would be frustrated at being unable to work in the field they studied, we can conclude that education and poverty are unrelated.

Unlike the "Latet" association, the agency officially empowered to determine the poverty rate in Israel is the National Insurance Institute, which defines a man as poor when his income is less than half the median income— exactly the accepted definition of relative poverty in the OECD states, in which relative poverty follows the Planck-Benford statistics—9%—while, according to the National Insurance Institute, the poverty rate in Israel is about 22%. Unsurprisingly, every year, in the days that follow the publication of the annual poverty report, every news report raises the severe problem of poverty in Israel, and commentators rush to name Israel the poorest nation on Earth! The National Insurance Institute does not base its data on surveys but on actual income tax reports. Supposedly this is more accurate, as you cannot cheat on tax reports—but of course you can. It all depends on how you count children. Radiation modes have no children and therefore, even though it may sound annoying, physically speaking one should not include children in this statistic, as they are unlinked nodes—they have no bank accounts). Economically speaking, though, children are an investment. One person decides to buy another apartment, and another uses the same amount of money to raise another child. The state also sees children as an investment encourages its citizenry to bring more and more children into the world. One day, these children will be the ones to maintain the economy, when they replace their parents who have become exhausted or passed on to the great beyond.

When the birth rate is higher than what is necessary to maintain the number of earners in a country, the population grows and drives the economy forward because of the need for more apartments, more roads, more communication resources, etc. The state, therefore, subsidizes the rearing of children. One way to encourage having children is to reduce income tax in proportion to the number of children the earner or earners have. This is accomplished using the concept of a "standard person." A single, unmarried earner is one standard person and pays his taxes in full. A family with children, on the other hand, comprises more persons and therefore has a smaller income per person; the smaller the income per person, the lower the income tax. In Europe, a "standard person" is calculated by the square root of the number of family members. For example, a family with two parents and seven children—nine persons in total—has three standard persons. In Israel, on the other hand, the National Insurance Institute employs a calculation method that makes a family with nine persons have 5.6 standard persons. Therefore, of those two families with the same income and the same number of children, the Israeli family will have a lower income-per-standard-person than the European one.

This shows how some people can specifically stress how bad the situation is. It is interesting that most commentators do not mention that poverty is calculated differently by different entities. As we mentioned earlier, due to the Israeli income support policy, it is advisable for the bottom decile to engage in jobs for which one can be paid black money. For example, the monthly income of a maid, who earns about NIS 50 per hour[49] of work and who works twenty hours a week, is earning NIS 4,000 that she does not report to the tax authorities. The National Insurance Institute sees this woman as an unemployed person, who is entitled to about NIS 2,000 per month in income support. She is further entitled to subsidized housing and other benefits. The bottom decile therefore comprises far fewer poor people than those cited in the National Insurance reports. As this phenomenon is well-known, someone had a brilliant idea and calculated the percentage of the poor based on their spending rather than their earnings. The poverty rate calculated thusly was found to be 8%, similar to the one suggested by the Planck-Benford distribution and the other OECD states.

---

49. 1 NIS is worth about $0.3.

Executive Salaries
The right person in the right place: how is a CEO selected, and why does he earn so much?

One of the most maddening things to see is a higher salary than your own. Despite the general disdain for high executive salaries in the general public, the salary gap remains enormous. For example, Wal-Mart's[50] CEO pocketed an annual salary of $19.4 million in 2015, about a thousand times the average salary in the company. It is not unreasonable to assume that in that same company, which has roughly 2.2 million employees, one could find a number of talented employees who would gladly do his work for an annual salary of only one million dollars—and might even do it just as well. One is reminded of the story of the penniless Jew who proposed a deal to baron Rothschild, which would have yielded a profit of one million gold coins to the latter: the poor Jew would marry Rothschild's daughter for a dowry of one million coins—one million fewer than the dowry proposed by Rothschild to the gentleman who was actually marrying his daughter.

How does Wal-Mart's board of directors reach a decision to pay a salary of $19.4 million to the CEO? In his book, "Parkinson's Law," Northcote Parkinson describes two methods used in appointing a navy officer out of a candidate pool: the English method and the Chinese method. The English method is based on an interview. The first question is, "Who is your father?" If the candidate says that his father is a shoemaker,

---

50. The largest retail company in the world.

the interview ends immediately. If the answer is, for example, "My father is an officer in Her Majesty's navy," the candidate passes to the second stage, in which he is asked, "Who is your grandfather?" and so on. In the Chinese method, which is equally effective according to Parkinson, all of the candidates are told to write an essay within a certain time period. The candidate who writes the longest essay gets the job. Obviously, Wal-Mart's CEO is not elected by either of these two methods. Wal-Mart's CEO is elected by the method that Napoleon used when appointing generals to his army. As the story goes, a pile of CVs from about forty generals lay on Napoleon's desk. Napoleon opened none of the folders; he pulled out one of them and said, "This is the man!" When asked how he chose a general for the army without opening even a single folder, he answered that he was looking for a lucky general. This imaginary story is probably based upon a question that Napoleon used to ask about any potential commander: "I know that he is a good general, but is he lucky?"

According to Benford's Law, the reason that the digit 9 appears less often than the digit 1 in corporate balance sheets is the following: if we liken a digit to a box and the value of the digit to the number of balls in the box, the digit 9 wins nine balls, compared to a single ball won by the digit 1. For the sake of precision, according to Planck-Benford's law it is 6.5 times harder to win nine balls than it is to win one. It is also much more difficult to be appointed CEO at Wal-Mart than to get a job as a warehouse worker in that same giant corporation.

Wal-Mart's CEO is not chosen by interviews, as derisively described by Parkinson, but is selected based on proven successes. We will generally see two types of CEOs: the entrepreneur CEO who started his own company, raised capital, and made the company successful while remaining its CEO. Such a case needs plenty of luck with its product, with the correct investors, and with the correct timing in the financial markets—preferably an economic boom. The other type is the CEO who was an employee and started as a salesman, as an engineer, or in the financial department, and was promoted higher and higher in the company following a series of successes. Senior employees often switch companies when they move to better positions due to the hierarchy that exists in every company; your promotion is highly dependent upon your boss. It is hard to see how a successful person in the financial department can approach his boss and argue for his own promotion if his boss has not managed to make progress in his own right. If the boss is not promoted, he cannot be expected to offer his own job to the talented young worker below him. If the boss is not a rising star himself, the talented employee would generally prefer to start working for another company in which he has better chances of being promoted. Nevertheless, the best way to get promoted in large organizations such as the military, the government, the academic world, and major corporations is to pick a talented boss and become his right hand. In this case, you avail yourself of the good luck of your direct manager. If the boss' career flourishes, you follow his footsteps and will soon have your own followers, just like in social media. We see such "dynasties" among the Nobel Prize winners, in the army, in industry, and in politics.

Success (i.e. the number of links) in politics, in the academic world, and in the military—and sometimes even in the arts—is more about prestige than money. Nowadays it is somewhat difficult to quantify prestige, but it is very likely that web-based search engines will soon allow us not only to find how many links a certain person has, but also the ratio between their positive and negative links and a weighted "prestige score" for that person. On Facebook, people tend to propose friendship to those who already have many friends, and to accept friendship proposals from such people. In science, new metrics are starting to emerge, such as the "impact factor," which represents a scientist's reputation numerically. This metric is governed by several criteria, such as number of scientific publications, the number of times they were quoted in others' work, and so forth.

We will now attempt to apply the Planck-Benford Law to an economic corporation. For the sake of simplicity, we will set aside promotion mechanisms that depend upon switching between companies and examine economic corporations in which salary amount is the metric of success.

For various reasons, which we will discuss later, the CEO's salary is expressed in comparison to the average salary. In the long tail distribution, the average salary is higher than the median salary.

One may show, using the Planck-Benford Law (Kafri & Fishof, 2016), that in a company with N employees, whose salaries are classifiable into N levels, the employee with the highest salary is in level $n=1$, the lowest-paid one is in level $n=N$, and the employee with the average salary is in level $n=\sqrt{N}-1+1$. For companies with more than a thousand employees, one may simplify things and say that the level of the average salary is $n\cong\sqrt{N}$. Data about the average salary level allows us to calculate the ratio between the CEO's salary (probably the highest) and the average salary. Since the normalization constant is equal for all salary levels, in a large company the ratio will be roughly $\ln 2/\ln(1+1/\sqrt{N})$. Following this formula, Wal-Mart's CEO can be expected to earn 1,034 times the average salary.

One may find the ratio between CEOs' salaries and the average salaries of the largest corporations in the world on the Internet. We examined those from the Fortune 100 list for 2015, which holds data pertaining to the world's 100 largest companies (Payscale, 2016). The company with the largest number of employees in the world—Wal-Mart—tops the list, and we found that the CEO's salary is 1,028 times the average salary in the company. 23 out of the 100 companies in the list at least approximately follow this rule-of-thumb formula. Five companies pay more than double the value derived from the formula, and five pay less than one-tenth that value.

If we add up the average of the CEO salaries of the 100 largest companies on Earth, we will see that the CEOs' salaries amount to 87% of the value predicted by the Planck-Benford Law. Let us not forget that within the 100 largest companies there are companies that intentionally avoid this model, the most striking example being Warren Buffett, who only draws an annual salary of one dollar.

The model we used to calculate the CEO's salary is based upon an extraordinarily simple assumption: we took N salaries and distributed the loot among them in a completely random way. We did not consider the tribulations undergone by the CEO as he moved between companies until he reached the top echelon of a large corporation. The story of the "successful CEO" is interesting and worthy of a biography; the molecule that traversed the greatest distance inside a gas balloon, compared to the other molecules, has a story about all of the collisions it underwent before reaching its outstanding accomplishment too. The story of the molecule, which really started from zero, will certainly rivet and touch the other molecules in the gas balloon, but the bottom line is that the molecules in the balloon follow the gas laws. Nobody believes that someone took the names of the employees in Wal-Mart, put them in a lottery machine of sorts, and appointed the winning number CEO. The equilibrium is dynamic. Even though the various echelons in the company are determined in a manner that maximizes entropy, the CEO also played a role in his own appointment: being in the right place at the right time, with the right personality and the right qualifications, similarly to Rabbi Akiva's maxim, "Everything is foreseen, yet free will is given"—in other words, the distribution is fixed but free will is given, and each individual has their own unique story.

In smaller companies, the level that holds the average salary is slightly different because the approximation we offered, $n \cong \sqrt{N}$, is inaccurate for small numbers. We will therefore present the ratio between the CEO's salary and the average salary in medium-sized and small companies, for which the approximation does not apply, in the following table:

No. of employees	CEO compensation relative to average salary
10	2
100	7
1,000	22
10,000	69

Table 2: The ratio between the CEO's salary and the average salary in medium-sized and small companies

Table 2 shows that, according to Planck-Benford, the salary of a CEO in a company with ten employees should be twice the average salary in the company. Conversely, in a company with a thousand employees, he will be paid about twenty times the average salary, and in a company with 10,000 employees, his wage will be about seventy times the average salary. As a matter of fact, table 2 is identical to table 1, the relative wealth of the deciles, except that table 2 is normalized to the average salary. In Israel, there is a law that limits the CEO's salary in financial institutions to not more than 35 times the lowest salary in the company. Let us now calculate CEO wages based on the minimum wage. Let us assume that the minimum wage belongs to the employee in level N, with N being the number of employees. Here we will not use the formula $\ln2/\ln(1+1/\sqrt{N})$, which expresses the highest salary as a function of the average one, but rather the formula $\ln2/\ln(1+1/N)$, which shows the ratio between the highest and the lowest salaries.

According to this formula, in a company with a statistical distribution, the CEO of a company with ten employees will receive about seven times the minimum wage. As of 2017, the minimum wage in Israel is about $1,250 per month, and therefore the CEO's salary will be about $8,750 per month. On the other hand, in a company with about a hundred employees, the CEO will receive $87,500 per month, and in a company with a thousand employees the CEO will receive close to $1 million—certainly an exaggerated salary.

On the other hand, if we look at table 2, which shows the CEO's salary as a function of the average salary in the company, and if we assume that the average salary is $2,500 per month, the CEO's salary in a company with ten employees would be $5,000, in a company with a hundred employees, $17,500, and in a company with a thousand employees, $55,000 per month—certainly more reasonable amounts.

No. of employees	CEO salary when the minimum salary is $1250/mo.	The ratio between the CEO salary and the minimum salary	CEO salary when the average salary is $2500/mo.	The ratio between the CEO salary and the average salary
10	$8,750	7	$5,000	2
100	$87,500	70	$17,500	7
1000	$886,867	709	$55,000	22

Table 3: The CEO's salary as a function of the average salary, compared to his salary as a function of the lowest salary, by number of employees.

What happened here? The simple answer is that this is the result of government regulation of the minimum wage. It is logical to set a minimum wage that would allow a worker to make a living, but the CEO's wage should not be tied to the minimum wage, as that would distort the probability-related equilibrium.

Economic regulation is mostly undertaken as a result of the democratic method. Since the poor always outnumber the rich, most voters are relatively poor, and therefore those wishing to be elected tend to promise to help the weaker strata break out of their plight, even though the distribution is fixed and unalterable. The simplest way of helping the weak is to financially help them, out of the taxpayers' money. As increased taxes are unpopular with the upper and middle classes, and as everyone understands the importance of these classes to the economy as they pay most of the taxes, the simplest solution is to provide help out of the pockets of employers, who are the minority, rather than from public funds. Raising the minimum wage paid by employers is a wonderful solution. Between us, the employers are well-to-do people, who force other people (who are in need) to get up in the morning and go to work....

For example, Election Day for the Knesset in Israel is a "holiday," similar to those in the rest of the democratic world. Employees have a day off from work on Election Day. Even though it is the employer who pays for this day, the employee does not have to prove to him that he has indeed exercised his right/duty to vote. Could anybody imagine the state allowing someone to exercise their lawful rights at the expense of the state, without torturing that person and demanding all manner of certifications?

Here, even though the act of voting generally takes only few minutes, the citizens receive a full day off, whether they vote or not. Why? Because the state is generous with other people's money. So long as the benefit given with the employers' money is good for the public, all is well. Between us, do you really care who pays for your free meal?

The problem is that, at the end of the day, excessive regulation worsens the economic conditions of the residents and is only good for the elected officials, who collect more votes. A good example of this is Venezuela under Hugo Chavez. Chavez championed an equal distribution of the capital and revenues from the nation's natural resources. He nationalized many private factories for the benefit of the public at large. As an example, here is the story of a small privately-owned plant, dedicated to extracting sugar from sugarcane. Before the Chavez era, the plant had about 350 employees who both grew the sugar canes and extracted sugar from them. Chavez believed in equality, so he nationalized the factory immediately upon winning the presidency. When the factory became government-owned, he could benefit both its employees and its new owners, i.e. "The People."

How? Chavez added posts to the nationalized company until it numbered about eight hundred employees. Such a measure is good for employment statistics, and indeed, the employment rate in Venezuela rose. Chavez also increased the salaries of those employees. Nowadays (2017), Venezuela, formerly the largest manufacturer of sugar in the world, has to import a portion of the sugar it consumes. Today, the factory, whose story we brought here, only employs 80 employees, who no longer grow the sugar cane but buy it abroad and only refine it. Since 1999, the year Chavez came to power, tens of thousands of hectares of sugarcane plantations were nationalized, as well as ten out of the sixteen private refineries. Since the economy of Venezuela is based on sugarcane and petroleum, Chavez financed his welfare programs by selling petroleum, an important natural resource in Venezuela. Once the prices of petroleum fell, the economy crashed. The result: nowadays (2017), about three years after Chavez's death, the economy in Venezuela is considered the worst in the world. The continued regulation of the prices of food and medications brought about shortages in these products because it meant importing them and selling them at a loss, which neither the importers nor the state could afford to do. As a result, imports stopped. Since 1999, Venezuela has lost 1.2 million production jobs, and out of 12,700 factories, only 4,000 remain. Hunger has become a serious problem in Venezuela in recent times.

Was Chavez a bad person? There is no physical definition of "bad." Chavez took the market out of equilibrium through extreme regulation and destroyed the economy. This is proof that we do not control the economy—the economy controls us.

Another example: the current minister of finance in Israel (2017) won his reputation through his dogged campaign against mobile phone providers, with the purpose of abolishing the exorbitant fees that they used to charge for any call between a subscriber of company A and a subscriber of company B. This fee was called a "connectivity fee." There is no question that the mobile phone providers should not have charged these high fees, especially as they enjoyed tremendous profitability even without the connectivity fee.

But money is all about flow. When a company amasses a large amount of money in its bank account, someone pounces and takes it over. How is this accomplished? The entity that wishes to take over the rich company borrows money from the public by issuing bonds. With the money raised through the public, that entity buys a sufficient number of the company's shares to control it. Then the entity withdraws funds from the treasury of the company, which are obviously profits, as a dividend. With these funds it pays its lawfully due taxes for having withdrawn dividends, repays its loan over several years, pays the principal and the interest to those who had invested in the bonds, pays interest to the banks, and makes a profit for himself. Everyone wins: the owners of the mobile phone provider who sold the company and made a profit, the banks who sold the bonds and received a handsome commission, the public that receives a better interest rate than it could get in the bank, and, of course, the company that has taken over another good, profitable company. In short, a great festival of money flows, where nobody loses.

The then Minister of Communications, who is the current Minister of Finance, had indeed managed to cancel the connectivity fee, but those connectivity fees would disappear anyway as technology kept advancing— nowadays it costs almost nothing to hold a conversation. In those days, when the Minister had just removed the connectivity fee and before the companies came to terms with the reduced fees, the mobile phone providers' funds ran out and the new company owners could not repay the bank. Those who lost the most in the wake of this very popular move were actually the investors who had purchased the bonds. Investors also lost due to the reduction in the value of shares, which followed the decline in the profitability of the companies. Most of the investors were the pension funds of the public, and the banks. As a result, the public, which was so glad to have its mobile phone bills reduced, will find itself with a smaller pension in the end. The great winner was the Minister of Communications, Mr. Robin Hood. And what befell the tycoon who tried to take over the mobile phone provider? He is in dire straits indeed. But let me tell you a secret: unfortunate though it may be, one cannot make a living out of schadenfreude. Otherwise, I would have been a billionaire myself!

The Distribution of Wealth—The Pareto Principle
Why are assets distributed unequally?

In the beginning of the twentieth century, the economy was much less developed than it is today. Most people literally worked for a living, unlike nowadays, when most people work to have a *better* living. The economic inequality at the time was not expressed as a gap between incomes, as incomes probably could not be calculated with any acceptable degree of precision back then; economic inequality was rather expressed as an unequal distribution of assets between people,[51] and we have seen that assets and money are different concepts. Nevertheless, since

---

51. Most of the population, except for government employees, did not receive fixed, established salaries, and therefore their income could not be precisely calculated.

the value of assets is the basis for the creation of money, it is worthwhile to examine the unequal distribution of assets in the context of the Planck-Benford Law.

The first person who inquired into the matter was Vilfredo Pareto, in 1879—two years before Newcomb published Benford's Law, about a similar distribution of digits. Pareto found that about twenty percent of the population held about eighty percent of the assets. This ratio, known as the Pareto Principle, is also called the 80:20 principle. This law is another one of those laws that are derived from the Planck-Benford Law. It is interesting to note that this law was first published by Planck, who published his theoretical work about the distribution of photon energy in modes in 1901. It seems that—with the exception of Zipf's Law, which was discovered in 1949 in the distribution of words in texts—all the laws derived from Planck's work were known for a very long time.

The name "80:20 principle" may mislead one into thinking that the sum of the percentage of owners of assets and the percentage of assets should be a hundred, even though the two quantities are unrelated. The digits on the right side represent the percentage, out of the population, that holds the percentage of assets on the left. It is therefore clear that this could have been the "65:20 principle" just as well, and it is coincidental that the sum of these two numbers is a hundred, but this name makes it easier to remember the principle. The Planck-Benford Law allows us to calculate what portion of assets is held by what portion of the population. It is of note that the Pareto Principle is not really a law, and does not give actual quantitative results. Indeed, the Pareto Principle is one that points out the unequal distribution of property.

Property is not money because it does not flow. It is, however, convertible into money under certain conditions. The care required in using the Planck-Benford Law in economics is seen when calculating the CEO's salary as a function of the minimum salary, as we showed in the previous section. A CEO salary that is a function of the minimum salary is much higher than actual CEO salaries, but the CEO's salary calculated as a function of the average salary matches actual salaries. This is because even though the relative distribution of wealth is independent of the general amount of wealth, it does depend upon the way we subdivide the population: into deciles or into percentiles. For example, the ratio between the top and bottom deciles is smaller than the ratio between the sum of the top ten percentiles and the sum of the bottom ten percentiles.

The reason for this is that we assume that everybody within a certain decile has equal incomes. Therefore, when we calculate the ratio between percentiles, we will notice both the richer and the poorer people, compared to the subdivision into deciles. The finer the subdivision, the greater the gaps.

Human poverty is limited both for regulatory reasons and for the simple reason that one needs a minimum amount of cash to survive, and therefore it follows that if someone is alive, they must have at least a bit of money. Compare this to the Planck-Benford Law, in which "poor" people, who do not have money to survive, exist.

In physics, when we measure the radiation of a black body, we cannot see the "poorest" photons in an energy-related sense, those whose wavelengths are longer than the Earth, and therefore they may as well not exist for us.[52]

---

52. We actually cannot see any photons whose wavelength is longer than the biggest antenna on Earth.

Those who study inequality measure CEO salaries as a function of the average wage, which is the correct procedure when working with the Planck-Benford Law. They also examine inequality in property by percentiles, not deciles. When we use the Planck-Benford Law we must take care to use it according to the economic data: where the data relate to percentiles, we have to use the Planck-Benford Law with N=100, meaning that the property of the highest decile will be calculated as the sum of the ten highest percentiles.

The OECD data for 2015 are calculated on a percentile basis, i.e. subdivision into percentiles. The theoretical calculations were therefore conducted on this basis as well.

Percentiles	Property % - OECD 2015	Calculated property %
1	18%	15%
10	50%	52%
20		66%
40	87%	80%

Table 4: The property percentage among population percentiles (Kafri & Fishof 2016).

In table 4, we compare the relative amounts of property owned by the various percentiles according to OECD statistics (in the middle column)[53] to the values obtained by the Planck-Benford Law (in the right column). The report does not tell us how much is owned by the top 20% of the population, but the answer should be ln 21/

53. Murtin & Mira, 2015

ln 101 = 0.66, according to Planck-Benford; 66% of the property is owned by 20% of the population. If we subdivide the population into thousandths, then according to the Planck-Benford Law the share of the top 20% will grow to 77%, similar to the original Pareto Principle.

We can see a good match between the Planck-Benford Law and the actual distribution of wealth, despite the problem of obtaining credible data about property rather than data about revenue.

Chapter 4

Boom and Bust

Why is it impossible to predict the arrival of a boom or a bust? Or: Why doesn't every question have an answer, even in mathematics?

Why are there booms and busts in economics? Why isn't the economy a calm bath? **The main reason is that the amount of money in the world is zero.** This means that booms occur when people take more loans from banks and do more business, and busts occur when people are afraid to take loans and therefore do less business. Booms and busts would not be possible were the amount of money in a given country finite. When loans are taken, money is generated; when they are discharged, money is nullified. When more loans are taken, more money flows, and when loans are repaid the amount of money in circulation decreases. There are always people who are afraid to take loans, and there will always be those who are always willing take the loan as long as someone is willing to lend them money.

According to the second law of thermodynamics, it may be statistically predicted that when two states are possible, pessimistic and optimistic, half of any group of people will be more pessimistic than average and half will be more optimistic. What would upset that balance? The answer is related to the thermodynamic characteristics of networks. Once there is a certain fluctuation in a balanced state, such as a burst of optimism with regard to a new technology or the discovery of a natural resource, such as a gas field, investors tend to join the optimistic wave and a consensus of optimism is created. After the prices skyrocket, some investors realize their gains, and decreases appear. At that point the other investors may become fearful and start selling excessively. In other words, when different opinions exist concerning investments, we tend to go by the majority opinion, as was already demonstrated in the scriptures—"Thou shalt not follow a multitude to do evil; neither shalt thou bear witness in a cause to turn aside after a multitude to pervert justice;"[54] and in the Latin maxim *Vox populi, vox Dei.*[55] The thermodynamic explanation for this is that a higher probability exists for a node to join a rich network than for it to join a poor one, in the same way that most people would prefer, according to Zipf's Law, to relocate to big cities. The tendency of the individual to accept the majority opinion is the source of economic crises and the reason for bubbles, such as the gold rush, or the craze for technology shares, or for real estate. Once the fickle majority opinion goes to the other extreme with regard to the value of assets that comprise an important part of the market, a certain type of economic crisis follows. The change usually starts with a mania that causes an exaggerated rise in the value of the assets, due to expectations that a further rise will

---

54. Exodus 23:2; It was already recognized back then that we would tend to follow a majority, even when we know that we are distorting justice—a strong human tendency.
55. The voice of the people is the voice of God.

occur. An important example is real estate: when there is a shortage of apartments in the market, their price rises and over-construction follows. After that there are too many apartments, and their price declines sharply. The problem becomes worse in cases of over-regulation and government intervention. This pattern also plays out when expectations that a certain technology will "change the world" lead to excessive investments in the stock exchange, when the fact is ignored that the processes governing the assimilation of a technology require decades—a much longer period than the time frame in which the investors expect to realize their investments. The first application that was spoken about when the laser was invented in the 1960s was communications through optical fibers. Nevertheless, even though the expectations were ultimately realized in every way, over fifty years have already passed and many copper cables are still alive and kicking, against the expectation that they would disappear within a decade.

Human behavior is viral. It is motivated by an impulse to imitate that owes its origin to the tendency of networks to grow. Consider the following story: the Tunisian Mohamed Bouazizi was a peddler. He had a vegetable cart, which he used to feed his family. Since he had problems with his license and did not pay a $280 fine (an amount equivalent to two months' salaries in Tunisia), he was caught by several policemen, his scales were confiscated, and he was thrown to the ground and beaten in front of passersby in the market. The humiliated Bouazizi went to complain in the city hall, where he was "thrown down the stairs," though not before being slapped by the clerk, to whom he complained in detail. Out of anger and in a burst of emotion, he bought a can of gasoline and set himself on fire. At age 26, Mohamed Bouazizi became a figure of historical importance: a day after the event, on December 17th, 2010, demonstrations broke out in Tunisia, leading to a change of government within less than a month. The riots spread out to Egypt, Yemen, Syria, Algeria, Jordan, Morocco, Iraq, Kuwait, and Sudan. In Syria and Yemen, a civil war is still being fought to this day (2017). In other countries, mass demonstrations occurred, with slogans of a similar nature. In that exact year, Israel also saw a huge public reaction, with about eight percent of the population taking to the streets in protest. Also in Israel, a man named Moshe Silman set himself on fire after suffering great injustices. The then-chairman of the National Union of Israeli Students, who was on his way to a trip abroad, returned immediately to Israel to lead the demonstrations started by a young, charismatic lady from a well-to-do family, who was complaining about the cost of living and high rental fees. People flowed out of the coffee shops in Tel Aviv and the development towns in the periphery, and threatened to change the government. The thing is that Israel is a democracy and one does not need a revolution to change the government. Indeed, new elections were held in January 2013 and the result was that the same prime minister was elected, rental fees kept rising, and two of the leaders of the social protests became new Knesset members. Even today people are still talking about the social protest, its causes, its motives, and its accomplishments. Shouldn't one consider the possibility that this may have been nothing but imitation?

In fact, human morals also originate from majority opinion. Is the majority right? The simplest answer is that if the majority thinks like I do, it is right, and if it does not think in the same way, it is wrong. There is no justice in nature; justice is a concept that varies between societies and times and is defined by the majority opinion of that place and in that time.

The surprising thing is that even though our society has a multitude of laws, restrictions and regulations, judges and legislators, cops and robbers—the Planck-Benford Law still works. What are all of its underlying assumptions? Each box has an equal probability of containing a ball, and the uncertainty in the system—called entropy—is at the highest level. It is hard for us to accept the existence of these assumptions, but the precision with which they apply in reality, not only to economics but to human behavior in general, is nothing short of astounding.

Why is this? Regarding money, the reason is that it behaves like heat, like flowing energy. In the subprime crisis of 2007, the banks sold the public the risk ingrained in the mortgages that were given to the bank, in the form of mortgage-backed securities. The assumption made by many was that the banks delivered a ticking time bomb of dangerous mortgages to the hands of the public. The banks failed to realize that they were in the same bath with the public— a bath resembling a black body—and thus, when the owners of real properties lose, the banks also lose, and finally, everybody loses. When tycoons go bankrupt, the entire public pays. In the same way that we have no control over economic crises, the influence of regulation is also very limited.

The false premise underlying the gas economy—that money is attached to merchants, as opposed to the heat economy, in which the money is traded between two merchants—leads to wrong conclusions, like those of Kahneman and Tversky, who reckoned that people hate losses more than they are eager to make gains. They derived their conclusions from surveys they conducted, in which various bets were offered to a sample of interviewees. They discovered that when "Alice" was offered a bet, in which there was a 50% risk for her to lose A dollars and a 50% chance to gain 2A dollars (double that amount), in most cases Alice rejected the offer, even though it was in her favor. Furthermore, they found that there was a chance that Alice would reject a bet which offered her a 50% chance to gain twenty times more (!) than she would risk.[56] Apparently, Alice sometimes rejects bets that are extremely biased in her favor, out of an irrational "hatred of loss."[57] Nevertheless, betting is an economic activity and all economic activity is a zero-sum game. The proposal to "Alice" should come from "Bob," not from a theoretical survey in which no real money changes hands. Bob must be a sworn aficionado of losses to propose bets of this sort. How much does Alice hate losses? Exactly as much as Bob loves them. It would therefore be wrong to conclude from the surveys that the hatred of losses is greater than the eagerness for profits. If Alice accepts Bob's offers, Bob will become poor and Alice will enrich herself. If Alice rejects Bob's offers, nothing will happen—nobody can lose money in a deal that they do not engage in.

---

56. The basic assumption is that A dollars is not a significant amount for Alice; otherwise her rejection of the bet offer is indeed economically logical.
57. By this logic, one may conclude that anyone who buys a lottery ticket is a lover of losses.

When we ban the use of drugs, the drugs do not disappear; a new market is created, a market that employs (among others) murderers, policemen, lawyers, judges, and wardens. Laws may come and go but money will forever flow. The uncertainty that drives the flow of money is not only an economic or a physical phenomenon; it also exists in the realms of the law, logics, and mathematics.

In 1971, Times magazine published a series of Pentagon documents on US involvement in Southeastern Asia. The documents were given to the magazine by a person who held them lawfully, but who was not authorized to deliver them to the paper. The Pentagon went to court, asking that the Times be ordered to stop further publication, which the court proceeded to order. The magazine appealed to the Supreme Court, and the Supreme Court overturned the earlier judgment with a majority of six against three, allowing the publication. One of the judges wrote that, "without a free and informed press, no smart and enlightened public opinion is possible." This is an extension of the concept of freedom of speech, taken from John Stuart Mill's principles. The original concept states that freedom of speech is a means for discovering "the truth."[58] Now the press has a right—according to

58. John Stuart Mill claimed that freedom of speech was a means for discovering the truth for four reasons: A) perhaps that, which is forbidden to say, is the truth; B) there are no absolute truths, and the more complete truth is obtained as a compromise between differing opinions that contain partial truths, and therefore all opinion must be heard and should not be limited; C) Truth itself gains strength as it combats lies. Therefore, even if the things that are forbidden to say are lies, they must be heard; D) Truth loses its convincing power when it had to be protected by force. Even if one defends the truth by limiting the freedom of thought and speech, this actually damages the truth.

the Supreme Court of the US—to reveal "confidential" facts that reached its hands illegally, as part of the right of the public to form informed opinions based on the leaked information. The situation after this precedent is vague; it creates problems in logic, which also exist in mathematics, providing binding precedents that justify criminal offenses (in this case, by delivering materials to those not entitled to receive them). This situation is similar to a finite system of statements that contradicts itself, such as the saying "this sentence is false." If the sentence is false, it is true; if it is true, it is false. Obviously, governments cannot afford the unrestrained distribution of confidential materials, and therefore in Israel a finer distinction was made such that, in the case of an "obvious and immediate danger," freedom of speech may be limited (Vegshel, 2014). Of course, this definition reveals the law for what it really is—a roulette wheel that depends upon the mood and political opinions of the judge, under the pressure of the media.

David Hilbert was one of the greatest mathematicians of the twentieth century. He belonged to a generation that was proficient in every area of mathematics. In 1895, he was appointed head of the faculty of mathematics in Gottingen University, the Mecca of mathematics in those days. Hilbert was among those who formulated the axioms governing Euclidean geometry, a partner to the formulation of General Relativity, and one of the fathers of Functional Analysis. In the beginning of the twentieth century, Hilbert organized a conference in which he delineated future directions for development in mathematics—directions which are still the beacon for its progress. In this conference, Hilbert presented

23 key problems in mathematics, as a challenge for the mathematicians of the twentieth century. Some of the problems that he presented are still unsolved. The second problem in his list was this: prove that the axiomatic system governing arithmetic is consistent.

In 1931, Kurt Gödel presented a proof showing that no axiomatic system in mathematics is complete and consistent, thus solving the second problem presented by Hilbert in 1900.

In 1928, about three years before Gödel presented his solution, Hilbert asked another question of equal importance: could one plan an algorithm that can give "true" or "false" as an answer when fed a mathematical or a logical statement? Alonzo Church presented a solution in 1936, and, independently of this, Alan Turing also presented a solution for the problem in 1937, which represented a breakthrough in computer science. The answer was a proof that no such general algorithm exists!

This presents a major problem: if it is true, then something basic goes against our intuition in both logic and in mathematics. There are statements in mathematics that can be proven to be true, there are statements that can be proven to be false, and there are statements that, by definition, one cannot know whether or not they are true. In any reality—economic, physical, or logical—uncertainty exists. A famous question in physics, one that demonstrates the principle of uncertainty, is called "Schrödinger's cat." Erwin Schrödinger, who was one of the fathers of quantum theory, invented this hypothetical experiment in order to demonstrate the uncertainty inherent in quantum states. We will modify it into a similar question, which we will call "Schrödinger's rooster." Suppose that a rooster is trapped inside a sealed box, with food and air that allow it to survive, but no way to know whether the rooster is dead or alive without opening the box. The lid of the box is made in such a way that the moment it opens, the rooster is killed. The question is: is the rooster dead or alive? In physics, the answer is that the rooster is in a quantum dead-alive state.

Schrödinger's rooster is an excellent example for economical investments. We cannot know what the value of our investments portfolio is until after we "kill the rooster," which is to say, after we realize the investments portfolio. Only then, after a pathological examination, can we know whether the rooster was dead or alive in the first place. After the rooster dies, the commentators appear and explain to us why the rooster was alive, or why it was dead. The commentators are always ready with explanations that match their views. This brings to mind the story of a scientist who asks his colleague to explain to him why A is greater than B. The colleague shuts himself in his room and returns with the question after a while: "A is greater than B because…" and here the first scientist cuts him short and says: "Hey, you didn't hear me correctly. I asked why B is greater than A," at which point the theoretician answers: "The explanation for this is even simpler."

Summary

The economy as an uncertainty

The film *Margin Call* describes the beginning of the financial crisis in 2007. A "margin call" is a term in economics, related to the securities that a person provides to a bank in order to receive a loan. The securities are assets that are pledged to the bank, and whose value rises and falls as the market fluctuates. If the value of the securities becomes smaller than the balance of the debt, the bank may ask the borrower to repay the difference between the amount of the debt and the value of the pledged asset, or to pledge additional assets. The difference is called a "margin," thus the wording for the term.

The great subprime crisis in 2007, from which the world has not yet recovered a decade later, started with a continuous rise in the prices of real properties that went on for years. Due to the natural tendency of investors to assume that things will continue the way they are and have been, banks in the US granted mortgages to home buyers, and were satisfied with having the home itself pledged. They required no further securities. When the market is rising this is acceptable; the danger in this type of lending policy comes when the prices of apartments fall sharply, and as a result the value of the apartment may suddenly be less than the value of the remaining mortgage. In this situation, the borrower would prefer to give the asset to the bank in most cases, rather than repay the loan. Since the banking business requires liquid money—unlike property rights, which require lengthy periods of time before they can be realized into cash, especially in a market with a surplus of housing (which is the exact situation when the value of apartments is declining)—the bank may easily become bankrupt. Instead of raising the bar and demanding more equity from apartment buyers—so that they could cover potential losses due to future decreases in the value of the asset—some "genius" came up with a brilliant idea: create a security in which these dangerous mortgages are "mixed" with other securities. Now the risk is being sold to the investor public as negotiable securities called "mortgage-backed securities." This pretty little idea both seemed to protect the banks from risk and made them rich, because even though they were selling poisonous high-risk securities, to the public it sounded as though the securities were backed up by something real, such as real estate (which was technically factually correct). All of these financial structures, which contain different risks from different markets, are too complex for the average human being to understand, and are calculated by first-class mathematicians with the help of powerful computers in order to prevent a catastrophe.

When the real estate market started to drop—and this always happens at some point—it turned out that in the formulas used to calculate risk, not every scenario was considered. The *Margin Call* film describes how one evening, an analyst finds out that the big investment bank in which he is working (probably a stand-in for Morgan Stanley Bank), has actually been bankrupt for about two weeks without anyone noticing. The employee informs his manager, and as a result the senior management of the bank holds a meeting in the middle of the night. The CEO decides to sell all of the mortgage-backed securities in the hands of the bank, at any price, before everybody knows about the catastrophe—even though the management of the bank knows full well that the securities are not worth even a fraction of their nominal value. The following day, the bank manages to sell its securities during a sharp price decrease, and the other banks join in for a panicked sale. The world crisis starts. The chief sales manager, named Sam Rogers in the film, feels guilty because he sold securities to clients who trusted his advice, with the clear knowledge that he would be causing them severe losses. He approaches the CEO and tells him that he wants to resign. In order to convince him to remain in the firm after the excellent job he did selling the securities to the bank's clients, the CEO makes the following speech:

"...if this is all for naught, then so is everything out there. It's just money, it's made up, a piece of paper with some pictures on it so we don't all kill each other trying to get something to eat. But it's not wrong and it's certainly not any different than it's ever been. Ever. 1637, 1797, 1819, `37, `57, `84, 1901, `07, 1929, `37, `73, 1987, 92, 97, 2000, and whatever this is gonna be called. They're just the same thing over and over.[59] We can't help ourselves, and you and I can't control it, stop it, slow it, or even ever so slightly alter it... We just react... and we get paid well for it if we're right... and get left by the side of the road if we're wrong. There's always been and there's always gonna be the same percentage of winners and losers, happy fucks and sad sacks, fat pigs and starving dogs in this world... yes there may be more of us today... but the percentages... they always stay exactly the same".

59. These are dates of crises in the world's economy.

The text, which was written by a Hollywood screenwriter, is amazingly factual. If we replace the term "money," what the CEO of the bank in the film calls "a piece of paper with some pictures on it," with the term "an entry in the bank registers," the bank CEO is saying that the ratio between income levels is fixed and that we cannot control the fluctuations of the economy, but are controlled by it. The Planck-Benford formula shows us why the ratio between the rich and the poor is fixed. In addition, since the formula is independent of the wealth of a nation, crises, booms, and busts are chaotic phenomena that behave like human mood swings. This begs the question, though, what is this formula good for? The answer is that the formula reflects a state of equilibrium, in which uncertainty is at its peak value. A state of equilibrium is a stable state. This formula allows us to see how far an economy is from equilibrium. If it is very far, one should expect it to change, for one reason or another, towards equilibrium. In a similar manner, one may check the salaries of executives in firms in light of this formula, and see whether they are underpaid or overpaid; one may predict the distribution of voters between parties in elections, and as we mentioned earlier, this law was used in a similar way by income tax investigators to check for falsehoods in financial reports. The formula says nothing about which party one should vote for or which is the optimum investments portfolio, even though we must remember that since economics mean uncertainty, even a risk-free investments portfolio is a risk in its own right....

Bibliography

Chandor., J. C. (2010). Margin Call. Retrieved from: http://www.imsdb.com/scripts/Margin-Call.html

Dragulescu, A.,Yakovenko,V. M. (2000). Statistical mechanics of money. *The European Physical Journal B*, *17*, 723-729.

Friedman, D. (2001). Drugs, violence and economics. Retrieved from: http://www.daviddfriedman.com/AcademicUBMUTKV/drugs_and_violence/Drugs_and_violence.html

Gillespie, P. (2016). 4 reasons why Venezuela became the world's worst economy. Retrieved from: http://money.cnn.com/2016/10/25/news/economy/venezuela-breaking-point

Kafri, O. (2014). Follow the multitude - A thermodynamic approach. *Natural Science*, *6*, 528-531. doi: 10.4236/ns.2014.67051.

Kafri, O. (2014a). Money, information and heat in social networks dynamics. *Mathematical Finance Letters, Article ID, 4.*

Kafri, O. (2016). A novel approach to probability. *Advances in Pure Mathematics*, *6*, 201-211. doi: 10.4236/apm.2016.64017

Kafri, O. (2016a). Economic inequality and the second law. *Journal of Economic and Social Thought, 3 (4)*, 476-480.

Kafri, O., & Fishof, E. (2016). Economic inequality as a statistical outcome. *Journal of Economic Bibliography*, *3(4)*, 570-576.

Kafri, O., & Kafri, H. (2013). Entropy - God's dice game. CreateSpace: Independent Publishing Platform.

Ksenzhek, O. (2007). *Money: Virtual energy.* Boca Raton, FL: Universal Publishers.

Murtin, F., & d'Ercole, M. M. (2015). Household wealth inequality across OECD countries: new OECD evidence. *OECD Statistic Brief* No.21. Retrieved from:https://www.oecd.org/std/household-wealth-inequality-across-OECD-countries-OECDSB21.pdf

PayScale, Inc. (2013). Human capital CEO pay: How much do CEOs make compared to their employees? Retrieved 16, June, 2016 from:http://www.payscale.com/data-packages/ceo-income-2013/fortune-100

Reuteman, R. (2010). The cost-and-benefit arguments around enforcement. Retrieved from: http://www.cnbc.com/id/36600923

Vagshal, H. (2001). Review on the subject: Freedom of the press, freedom of opinion and national security. *Ministry of Public Security, Jerusalem, Israel.* (Hebrew), Retrieved from: http://archive.mops.gov.il/documents/publications/informationcenter/skirotmikzoiyot/pressandpublicsecurity.pdf